SpringerBriefs in Molecular Science

For further volumes:
http://www.springer.com/series/8898

Bang-Ce Ye · Min Zhang
Bin-Cheng Yin

Nano-Bio Probe Design and Its Application for Biochemical Analysis

 Springer

Bang-Ce Ye
State Key Laboratory of Bioreactor
 Engineering
East China University of Science
 and Technology
Shanghai
China

Bin-Cheng Yin
State Key Laboratory of Bioreactor
 Engineering
East China University of Science
 and Technology
Shanghai
China

Min Zhang
State Key Laboratory of Bioreactor
 Engineering
East China University of Science
 and Technology
Shanghai
China

ISSN 2191-5407   ISSN 2191-5415 (electronic)
ISBN 978-3-642-29542-3   ISBN 978-3-642-29543-0 (eBook)
DOI 10.1007/978-3-642-29543-0
Springer Heidelberg New York Dordrecht London

Library of Congress Control Number: 2012936962

Preface

Over the past decades, nanomaterials are at the leading edge of the rapidly developing field of nanotechnology both in scientific knowledge and in commercial applications. Their unique size-dependent properties make these materials have enormous potential as novel probe candidates in biochemical analysis. Therefore, different types of nanomaterials have been engineered to assemble biomolecules to construct novel molecular probes, named as nano-bio probe. Nano-bio probe combines the advanages both: the specific recognition and catalytic properties of biomaterials and the unique optical, electronic, thermal, and catalytic functions of nanomaterials. In recent years, the nano-bio probe is becoming increasingly popular as a routine bioassay, which could have comparable or even better sensitivity and selectivity to conventional molecular probe, not only for developing analytical and monitoring devices, but also for molecular-scale bioreactors. As expected, it has extended the limits of current molecular probes in biochemical analysis. We wrote this book with two goals: first to provide the reader with a brief overview of nano-bio probe design and application from different aspects, and second to try to introduce our group's work in this exciting field. We wish this book will be helpful to those who with closely related with research interests.

Bang-Ce Ye
Min Zhang
Bin-Cheng Yin

Contents

Chapter 1
Introduction

Over the past decades, the field of biochemical analysis has witnessed an explosion of interest in the employment of nanomaterials in various applications, such as environmental monitoring, developmental biology, clinical toxicology, and industrial process monitoring. In recent years, there has been an increasing interest in developing more robust, practical, and highly sensitive and selective detection assays that can address the disadvantages of conventional technologies. Consequently it employs nanomaterials to construct novel tools for studying biosystems, and researchers are investing a great deal of time and effort in developing new nanomaterials for applications in biodetection.

Nanomaterials have been the subject of enormous interest and are at the leading edge of the rapidly developing field of nanotechnology. The prefix "nano", is derived from the Greek word "nanos", signifying dwarf. Today this prefix is used to describe one billionth of a meter (10^{-9}) in the metric system. Nanomaterials are materials that are in a very small feature size in the range of 1–100 nm in at least one dimension. Due to the extremely small feature size and correspondingly large surface-to-volume ratio, nanomaterials have the potential for a wide variety of applications; therefore have obtained large amounts of funding from private enterprises and government. Most-used nanomaterials can be organized into five types: metal-based materials, carbon-based materials, polymeric materials, dendrimers, and composite materials. The unique properties of these various types of nanomaterials give them novel optical, catalytic, electrical, magnetic, thermal, mechanical, or imaging features. These features make nanomaterials superior and indispensable in developing novel probes for biological applications.

With the rapid development of molecular nanotechnology, engineered nanomaterials have merged into biological systems. Biomolecules such as DNA, RNA, enzymes, peptides, and antigens/antibodies, possess nanometric sizes comparable to nanomaterials such as nanoparticles, nanotubes, nanorods, quantum dots or nanowires. Via the assembly of biomolecules with nanomaterials, we could obtain a new type of probe: nano-bio probes, which combine the specific

B.-C. Ye et al., *Nano-Bio Probe Design and Its Application for Biochemical Analysis*,
SpringerBriefs in Molecular Science, DOI: 10.1007/978-3-642-29543-0_1,
© The Author(s) 2012

recognition and catalytic properties of biomaterials with the unique optical, electronic, thermal, and catalytic functions of nanomaterials. Undoubtedly, the nano-bio probe will overpass the limits of current molecular probes in biochemical analysis. Especially, nano-bio probe will play an important role in the development of molecular diagnostics, therapeutics, bioengineering, and molecular biology.

This book focuses on the recent progress and our group's work in nano-bio probe design and its applications for biochemical analysis, which is divided into four parts. The first part introduces a novel nanomaterial-enhanced fluorescence polarization assay; compared to the traditional fluorescence polarization assay, its performance and sensitivity are greatly improved. The applications of this novel assay in metal ions detection and protease assays are also presented. The second part introduces a novel two-dimensional nanomaterial: graphene (2010 Nobel Prize in Physics), which has opened up an exciting new field in the science and technology of nanomaterials. A fascinating finding is that graphene oxide (GO) can quench the fluorescence of dyes or chromophores via resonance energy transfer when in close proximity. Several GO-based biosensors have been developed by taking advantage of the strong adsorption of dye-labeled ssDNA or peptides onto the GO surface, and subsequent efficient fluorescence quenching of the dyes. These biosensors have been successfully applied to detect ssDNA, metal ions, helices, small molecules, insulin, miRNAs, proteases, pathogens, and living cells. The third part reviews the recent development of a colorimetric assay utilizing metal nanoparticles (MNPs) as sensing elements based on their unique surface plasmon resonance properties. Moreover, we will introduce our group's work in designing novel MNPs-based colorimetric probes for the detection of biological thiols, chiral recognition of enantiomers, visualization of phthalates and multiplex detection of metal ions. The fourth part gives a general overview on the recent development of metal-nanocluster-based luminescent probe design and its applications in detecting various targets (DNA, miRNAs and metal ions), and in cellular labeling or imaging, based on their excellent photostability and subnanometre size.

Chapter 2
Nanomaterial-Enhanced Fluorescence Polarization and Its Application

2.1 Principles of Fluorescence Polarization

Fluorescence polarization (FP) was first applied in biochemistry in the 1950s, when Gregorio Weber reported his studies on bovine serum albumin and ovalbumin conjugated with 1-dimethylaminonaphthalene-5-sulfonyl chloride (dansyl chloride) [1]. FP methods have become increasingly popular following Weber's work during the past few decades. The increase in the number and diversity of fluorescence polarization studies has been astonishing and the method is an intrinsically powerful technique for the rapid and homogeneous analysis of molecular interactions in biological/chemical systems. The technique has been successfully employed in many diagnostic fields including the detection and quantitation of antigens or antibodies based upon antibody–antigen interactions, protein-DNA interactions, protein-RNA interactions, DNA detection and the monitoring of therapeutic drug levels. It has also been used in monitoring enzyme-catalyzed hydrolysis, protein–protein interactions, and high-throughput screening during the course of drug discovery.

The fluorophore rotation can be calculated in a more quantitative manner [2]. Depolarization of the emission occurs if the fluorophore, and hence the emission dipole, rotates through an angle ω between the moment of excitation and emission. In fact,

$$\frac{1}{P} - \frac{1}{3} = \left(\frac{1}{P_0} - \frac{1}{3} \right) \left(\frac{2}{3 \cos^2 \omega - 1} \right) \tag{2.1}$$

where P is the observed polarization. The total depolarization is determined by an intrinsic factor (P_0) and an extrinsic factor (ω). Perrin related the observed polarization to the excited state lifetime and the rotational diffusion of a fluorophore to derive the famous equation [3],

B.-C. Ye et al., *Nano-Bio Probe Design and Its Application for Biochemical Analysis*, SpringerBriefs in Molecular Science, DOI: 10.1007/978-3-642-29543-0_2, © The Author(s) 2012

$$\frac{1}{P} - \frac{1}{3} = \left(\frac{1}{P_0} - \frac{1}{3}\right)\left(1 + \frac{RT}{\eta T}\tau\right) \tag{2.2}$$

where P is the observed polarization, P_0 is the limiting or intrinsic polarization, V is the effective molar volume of the rotating unit, R is the universal gas constant, T is the absolute temperature, η is the viscosity, and τ is the excited state lifetime.

For a spherical molecule, it follows that,

$$\rho_0 = \frac{3\eta V}{RT} \tag{2.3}$$

For a spherical protein, it follows that,

$$\rho_0 = \frac{3\eta M(\upsilon + h)}{RT} \tag{2.4}$$

where M is the molecular weight, υ is the partial specific volume, and h is the degree of hydration.

Combined with the Eq. 2.1, the Eq. 2.2 can be rewritten as follows,

$$\frac{1}{P} - \frac{1}{3} = \left(\frac{1}{P_0} - \frac{1}{3}\right)\left(1 + \frac{3\tau}{\rho}\right) \tag{2.5}$$

where ρ is the Debye rotational relaxation time, which is the time for a given orientation to rotate through an angle given by the *arccos* e^{-1}, which is 68.42°.

From Perrin equation (2.5), one can obtain that the FP value P is sensitive to a change in any of the three parameters ρ, τ, or P_0. Experimentally, τ is depending on the probe environment, with changes often associated with a shift from a hydrophilic to hydrophobic environment, or vice versa. Changes in the P_0 at a fixed excitation wavelength are uncommon, but changes in the effective P_0 can occur when energy transfer is present, resulting in a measured P_0 that is lower than the true P_0. Therefore, the FP value of a fluorophore is proportional to its rotational relaxation time, which in turn depends upon its molecular volume (molecular weight) at the constant temperature and solution viscosity. In a system where a random array of molecules is freely rotating, the FP value will be between 0 and 500 mP depending upon the rotation rate of the molecules. If a molecule is small, it will rotate faster and hence will have a smaller *FP* value. Conversely, the larger molecules will have bigger *FP* values due to their slow rotation.

Another term frequently used in the context of polarized emission is anisotropy, designated as r, which is calculated as

$$r = \frac{2P}{3 - P} \tag{2.6}$$

By analogy to polarization, the limits of anisotropy in completely oriented systems are +1 to −0.5 [2].

2.2 Sensitive Detection of Hg^{2+} by Fluorescence Polarization Enhanced by Gold Nanoparticle

Colloidal gold is a suspension (or colloid) of sub-micrometer-sized particles of gold in water, usually. The liquid displays either an intense red color (for particles less than 100 nm), or a dirty yellowish color (for larger particles). Gold nanoparticles (AuNPs) are colloidal gold particles with dimensions ranging from 1 to 100 nm. AuNPs possess unique optical, electronic, catalytic, and molecular-recognition properties, which make them as the research subject in a wide variety of areas, including electron microscopy, electronics, nanotechnology, and materials science. Especially, enhancement functions of AuNPs have been employed for signal generation and transduction in developing novel biosensing systems, in order to substantially improve the performance and sensitivity. For example, the enhanced effects were implemented by the use of AuNPs as labels for the amplified quartz crystal microbalance (QCM) detection and electrochemical detection. AuNPs can be used as electron relays to facilitate the interfacial electron transfer on electrode, or as fluorescence enhancer and quencher. AuNP-enhanced effects in a responsive polymer gel and in network field-effect transistors have also been reported.

In the published papers, the amplified signals and improved sensitivity of the different biosensing systems by incorporating AuNPs have been revealed. For example, AuNPs were employed to enhance microgravimetric quartz-crystal-microbalance measurements. Willner's group developed a novel method, based on the AuNP-catalyzed deposition of gold, for the amplified microgravimetric QCM detection of DNA [4] and thrombin [5]. The detection sensitivity of QCM for DNA can be significantly improved to 10^{-15} M by using AuNPs as "seeding catalyst" for gold precipitation on the transducer. Yan et al. [6] described the systematic evolution of improved electrical contacting between glucose oxidase and the electrode by the biocatalytic enlargement of AuNPs tethered to the enzyme electrode. They demonstrated that the enlargement of AuNPs facilitated the interfacial electron transfer, thereby enhanced the mediated bioelectrocatalyzed oxidation of glucose, and substantially improved the bioelectrocatalytic perfor-mance of the enzyme electrode. Dong et al. [7] reported that the detection sensitivity of single-walled carbon nanotube network field-effect transistors (SNFETs) for DNA can be further improved to ca. 100 fM by using AuNPs, in which the target DNAs were hybridized with probe DNAs on the device, and reporter DNAs labeled with AuNPs flank a segment of the target DNA sequence. Fan et al. [8] developed a "sandwich-type" detection strategy for sequence-specific detection of femtomolar DNA via a AuNPs-mediated amplified chronocoulometric DNA sensor. Minko et al. [9] investigated localized plasmon resonance excited in AuNPs coupled with a responsive polymer gel. The results demonstrated that a specially designed structure (vertically aligned cylindrical pores decorated with gold nanoparticles) of responsive polymer gel thin films were used as a platform for the transduction of external signals/stimuli into a strong optical effect that is enhanced by interactions of AuNPs.

Contamination of the environment with heavy metal ions has been an important concern throughout the world for decades. Mercury accumulated in vital organs and tissues, such as the liver, brain, and heart muscle, is a highly toxic material with lethal effects on living systems. Mercury originates mainly from coal-burning power plants, oceanic and volcanic emissions, gold mining and combustion of waste [10]. Also, microbial biomethylation of Hg^{2+} yields methyl mercury, a potent neurotoxin that concentrates through the food chain in the tissues of fish and marine mammals [11, 12]. Therefore, it should be highly desirable to develop a sensitive and selective mercury detection method that can provide simple, practical, and high-throughput routine determination of Hg^{2+} level for both environmental and food samples. Much effort has been devoted toward the design of Hg^{2+} sensing systems, including sensors based on organic chromophores [13–17] or fluorophores [18–21], conjugated polymers [22], DNAzyme [23], gold nanoparticles [24], semiconductor quantum dots [25], proteins [26], and genetically engineered bacteria [27]. However, most of these methods have either limitations with respect to poor selectivity with interference from closely related metals, or insufficient sensitivity [limit of detection (LOD) > 100 nM], and in certain cases not stable or functional in aqueous media due to low water solubility.

Another emerging approach for the detection of Hg^{2+} ion involves the use of oligonucleotide. Recent studies have demonstrated that Hg^{2+} ions can specifically interact with thymine bases to form strong and stable thymine-Hg^{2+}-thymine complexes (T-Hg^{2+}-T). Various Hg^{2+} ion detection assays based on this property of T-Hg^{2+}-T coordination chemistry have been developed in recent years [28–31]. Ono and Togashi [28] described the simple method based on the Hg^{2+}-induced formation of DNA folding that yields an intramolecular fluorescence resonance energy transfer process to detect Hg^{2+} with high selectivity and sensitivity (up to 40 nM). Mirkin et al. [29]-recently reported the colorimetric detection of Hg^{2+} in aqueous media using DNA-functionalized gold nanoparticle probes with deliberately designed T–T mismatches (sensitivity up to 100 nM). Further, Liu et al. [30] developed a one-step, room temperature, colorimetric assay of Hg^{2+} using DNA/nanoparticle conjugates (sensitivity of 1 µM). According to the U.S. Environmental Protection Agency (EPA) standard, the maximum allowable level (MAL) of Hg^{2+} in drinking water is 10 nM (2.0 ppb). This concentration is much lower than the detection limit of most available assays. Thus, the development of facile and practical assay for Hg^{2+} detection with higher sensitivity remains a challenge.

In this section, we apply fluorescence polarization assay (FPA) to analyze Hg^{2+} ion using a "AuNPs enhancement" approach [32]. The large surface area of AuNPs can increase the amount of assembled molecules, effectively raise the complex size of recognition molecule and target molecule, thus reach the aim of signal amplification, which significantly improve the detection sensitivity and stability. Our method to analyze Hg^{2+} ion is based on the formation of T-Hg^{2+}-T complexes. The detection sensitivity can be significantly improved to 0.2 ppb (1.0 nM) by using the "AuNP enhancement" approach. The outline of the method is shown in Fig. 2.1. Two complementary probes (Probe A, AuNP-5′S-A_{10}-GCTTCTGTTCTCT3′; and Probe B, 5′FAM-GTTGTGTTCAGTTGC3′), which

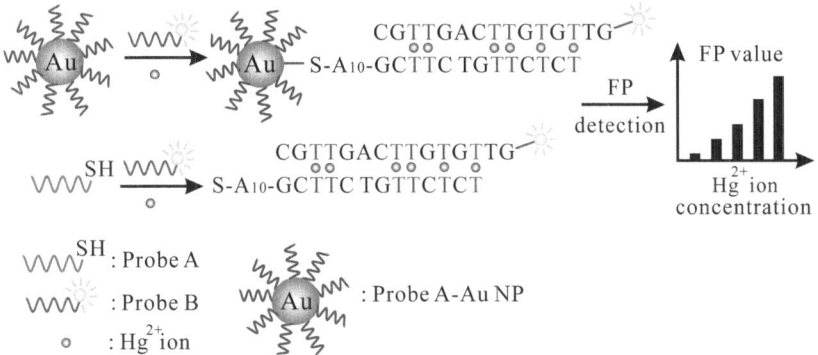

Fig. 2.1 Schematic illustration of the strategy for Hg^{2+} ion detection using fluorescence polarization via gold nanoparticle enhancement. Reproduced from Ref. [32], with the permission of Wiley John and Sons

contain strategically placed six thymine–thymine mismatches complexed with Hg^{2+} ions, were prepared. Probe A was modified with AuNPs (13 nm in diameter). Probe B was labeled with the fluorescence dye, 6-fluorescein-CE phosphoramidite (FAM). The stable hybridization between Probe A and Probe B occurs only when Hg^{2+} ion is present. From Perrin equation (2.5), the *FP* value will significantly increase when a fluorophore-labeled Probe B binds to AuNPs surface in the presence of Hg^{2+} ions.

AuNPs may enhance or quench the dye fluorescence, depending upon the size and the precise distance between the dye and AuNP [33–35]. At the small distances (1–3 nm), AuNPs can serve as ultra-efficient quenchers of molecular excitation energy in fluorophore-AuNP composites [36–39], outranging the quenching efficiency of organic acceptor molecules. Schneider et al. [40] employed the layer-by-layer deposition of oppositely charged polyelectrolytes onto 13 nm-diameter gold colloids to fabricate metal core-polymer shell capsules, in which fluorescent organic dyes fluorescein and lassamine rhodamine B were situated at various distances from the gold core. They found that the gold nanocore quenched the fluorescence of the fluorescein and lassamine dyes significantly with 20 spacer layers. Thus the large fluorescence quenching efficiency of AuNPs (QE > 98%) has been employed for DNA detection. Li and Rothberg [39] reported a sensitive detection assay of oligonucleotide sequences using the different electrostatic properties of single- and double-stranded DNA (ss-DNA and ds-DNA) to absorb on citrate-coated gold nanoparticles. The unhybridized oligos labeled with fluorescence would efficiently adsorb onto the gold nanoparticles so that their fluorescence was quenched. Dubertret et al. [36] described a novel molecular beacon using a hybrid material composed of 1.4 nm diameter gold nanoparticle, an organic dye, and a 25 bp ss-DNA stem-loop molecule. That was, a nucleic acid probe with-hairpin shaped structure in which the 5′ and 3′ ends are self-complementary, bringing a fluorophore and a nanoparticle into close proximity to quench fluorescence. The results demonstrated that this composite

Table 2.1 Samples used in this experiment

Samples	Components (Final conc.)
#1	FAM-Probe B (10 nM)
#2	FAM-Probe B (10 nM) + AuNP-Probe A[a] (10 nM)
#3	FAM-Probe B (10 nM) + AuNP-Probe A (10 nM) + Hg^{2+} (10 nM)
#4	FAM-Probe B (10 nM) + SH-Probe A (10 nM)
#5	FAM-Probe B (10 nM) + SH-Probe A (10 nM) + Hg^{2+} (10 nM)
#6	FAM-Probe B (10 nM) + TAMRA-Probe A (10 nM)
#7	FAM-Probe B (10 nM) + TAMRA-Probe A (10 nM) + Hg^{2+} (10 nM)

[a] The concentration of AuNP-Probe A represented the oligo concentration, which was 10 nM; while the AuNPs concentration was 0.1 nM

molecule was a different type of molecular beacon with a sensitivity enhanced up to 100-fold. However, in those cases where the fluorophores and the AuNPs are separated by bulky spacers, such as antibodies and DNA, the fluorescence is less quenched. Moreover, enhanced fluorescence is frequently reported if the fluorophores are placed within 3–50 nm above a surface covered with nanostructured noble metals [41–45]. The effect has been called metal-enhanced fluorescence (MEF) or surface-enhanced fluorescence (SEF).

Considering the fluorescence quenching properties of AuNPs, the case that AuNPs would quench the fluorescence of FAM-labeled Probe B (FAM-Probe B, 5′-FAM-GTTGTGTTCAGTTGC-3′) and then result in non-uniformity of the assay, would probably occur. Therefore it is necessary to study the effects of AuNPs and molecular quencher (TAMRA) on fluorescence intensity (FI) and *FP* values in the biosensing system. In this study, we chose molecular quencher (TAMRA) to modify Probe A (TAMRA-Probe A, 5′-TAMRA-A_{10}-GCTTCTGTTCTCTAC-3′) to carry out the molecular quencher experiment among DNA–DNA systems to compare the quenching effect performance between the nanoparticles and quencher. The thiol-modified oligonucleotides (SH-Probe A, 5′-SH-A_{10}-GCTTCTGTTCTCTAC-3′) were used to functionalize AuNPs (13 nm in diameter) to obtain AuNP-modified DNA (AuNP-Probe A, 5′-AuNP-SH-A_{10}-GCTTCTGTTCTCTAC-3′). All measurements of fluorescence intensities and FP values were done in the same condition. Seven samples were prepared in 0.1 M $NaNO_3$/10 mM MOPS (pH 7.5) as listed in Table 2.1.

As shown in Fig. 2.2a, Samples #1, #2, #3, #4, and #5 had the similar FI (F_1, corresponding to FI of free FAM-Probe B of 10 nM), while the FI of Samples #6 and #7 was a little lower than F_1 due to the quenching property of TAMRA. When detecting the FP, there were almost the same FP values (FP_1, corresponding to FP value of free FAM-Probe B of 10 nM) among the samples #1, #2, #4, #5, #6 and #7, there was increased FP in sample #3 (a substantial increasing because of AuNP enhancement (Fig. 2.2b). However, the reasonable polarization changes of Samples #5 and #7 to Samples #4 and #6 were not observed at a low concentration of 10 nM Hg^{2+}. This is because FPA has insufficient sensitivity without AuNP enhancement. These experiment results demonstrated that the gold nanoparticle has no quenching effect on the FI of FAM-labeled Probe B.

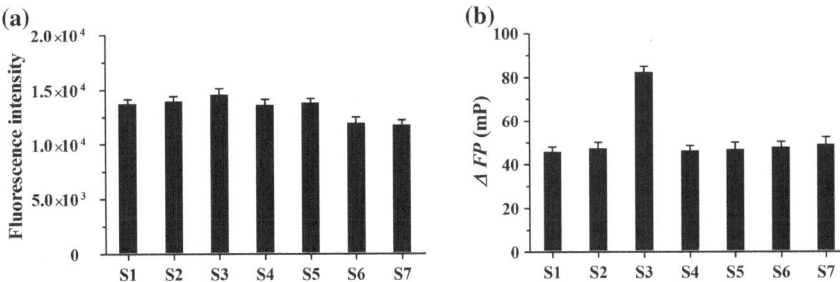

Fig. 2.2 The effects of AuNPs and TAMRA on the fluorescence intensity and *FP* value of dye-labeled Probe B. Reproduced from Ref. [32], with the permission of Wiley John and Sons

In the study conducted by Malicka et al. [43], a dense matrix of data for the dye-DNA-AuNP hybrid diameters versus ss-DNA length has been provided for many different surface coverages. The extrapolation of these data allows us to estimate the diameters of Probe A-AuNP hybrids used in this work. From the hybrid diameters $h \approx 32$ nm (oligo of 23 bp) we can derive the distance $d = (h-13 \text{ nm})/2 = 9.5$ nm between the 3′-end part of Probe A and the surface of the AuNPs. The effective diameter depends on the conformation of oligos adsorbed on the AuNP. The stable hybridization of Probe A on the surface of the AuNP with FAM-labeled Probe B occurs when Hg^{2+} ion is present. For high surface coverage and ds-DNA, the DNA would be shown to be oriented perpendicular to the surface and fully stretched. The diameter between the conjugated fluorophoric part of the FAM molecules and the surface of the AuNPs should be larger than 9.5 nm. Our experimental results showed that the fluorescence was less quenched by AuNPs in the case where the fluorophores and the AuNPs were separated at above 9.5 nm by DNA oligos. Therefore, we believe the fluorescence quenching or enhancing properties of the gold nanoparticles have no effect on the AuNP-enhanced FPA of Hg^{2+} in this work.

To evaluate the sensitivity of the FPA enhanced by AuNPs, different concentrations of Hg^{2+} from one stock solution were tested. The *mP* value was sensitive to Hg^{2+} ions and increased as Hg^{2+} increased (Fig. 2.3). The detection limit of this method is approximately 1.0 nM (0.2 ppb) Hg^{2+}. It is lower than the MAL of Hg^{2+} ion in the drinking water of 10 nM (2 ppb) in the U.S. EPA. The result illustrated that the method has a wide range of detection of Hg^{2+} ion from 1.0 nM to 1.0 mM. The selectivity of the method was also investigated by testing the response of the assay to other metal ions, including Mg^{2+}, Pb^{2+}, Cd^{2+}, Mg^{2+}, Ca^{2+}, K$^+$, Na$^+$, Zn^{2+}, Fe^{2+} at a concentration of 10.0 μM, which is 1,000-fold in excess as compared to that for Hg^{2+}. The results demonstrated excellent selectivity over alkali, alkaline earth, and transition heavy metal ions. There was very little *mP* value increase observed in the presence of other metal ions. The specific detection for Hg^{2+} is mainly attributed to its chelating ability of T–T mismatches, resulting in the formation of stable T-Hg^{2+}-T complexes.

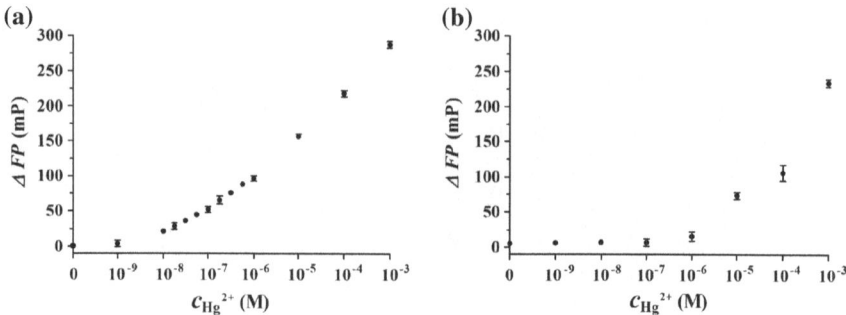

Fig. 2.3 Plots of fluorescence polarization changes as function of Hg^{2+} ion concentrations (*left*: AuNP-DNA; *right*: DNA–DNA). Reproduced from Ref. [32], with the permission of Wiley John and Sons

To assess the applicability of the method, the real samples and the spiked samples of river water were analyzed at two different concentrations (0.4 and 4 μM). The method revealed the good recovery rates from 97.6 to 102.5%. No interference in river water was observed.

In summary, we developed a highly sensitive and selective Hg^{2+} determination method at room temperature using thymine-Hg^{2+}-thymine coordination chemistry and FP via gold nanoparticle enhancement. Our method demonstrated several analytical advantages. First, it has high sensitivity with 0.2 ppb, which can be 2–3 orders of magnitude more sensitive than other techniques. Second, it is highly selective, which allows detection of Hg^{2+} in the presence of an excess (1,000-fold) of other metal ions in samples. Third, it takes only approximately 10 min to determine the concentration of mercury in aqueous media. Last, the assay can be carried out in 96- or 384-well plates, can render it suitable for routine high-throughput applications. We believe that the method has enormous potential for application of mercury monitoring in environment, water, and food samples.

2.3 DNAzyme-Self-Assembled Gold Nanoparticles for Determination of Cu^{2+} and Pb^{2+} Using Fluorescence Polarization Assay

For a long time it was believed that all enzymes are proteins and this thought has been changed with the discovery of the first natural ribozymes by Cech and Altman [46, 47]. A ribozyme is an RNA molecule with a well defined tertiary structure that enables it to catalyze chemical reactions, including RNA cleavage [48], splicing [49], and peptide bond formation in ribosomes [50]. The first deoxyribozyme was discovered in 1994-by Breaker [51]. Since then, intensive research has focused on the identification of the new nucleic acid enzymes, elucidation of mechanism of catalytic activity, and development of pratical application. So far, all known

DNAzymes, obtained by a combinatorial method called in vitro selection or systematic evolution of ligands by exponential enrichment (SELEX) [51–55], are not only the static depository of genetic information, but also can catalyze a surprising variety of biochemical reactions, including RNAcleavage, DNA cleavage, RNA ligation, DNA ligation, RNA branching and lariat RNA formation, DNA phosphorylation, capping, deglycosylation, cleavage of the phosphoramidate bond, photocleavage of thymine dimers, porphyrin metalation, and other enzymatic activities as peroxidases [56]. DNAzymes are preferentially selected to develop optical or electronic sensor due to their impressive chemical stability compared with RNAzymes and other enzymes.

Similar to protein enzymes, one kind of DNAzyme formed by G-quadruplexes (also called G-quadruplex-based DNAzymes), is capable of recruiting hemin as a cofactor, mimics horseradish peroxidase (HRP) and catalyzes the H_2O_2-mediated oxidation of 2,2-azinobis(3-ethylbenzothiazoline)-6-sulfonic acid (ABTS) or luminol. The G-quadruplex-based DNAzyme binding with hemin has peroxidase-like activity and has been extensively studied as colorimetric and/or chemiluminescence biosensors for the sensitive and specific detection of proteins, metal ions, and other biomolecules, indicating a great potential in biological applications. Willner's group pioneered the application of this peroxidase DNAzyme for signal amplification in analytical applications. For example, they developed DNAzyme-functionalized AuNPs as catalytic labels for the generation of biochemiluminescence [57]. The system was applied to detect DNA or telomerase activity. The DNAzyme stimulates, in the presence of hemin, H_2O_2, and luminol, the generation of chemiluminescence. The method had a detection limit for the detection of DNA of 1×10^{-10} M, and enabled the detection of telomerase activity originating from 1,000 HeLa cells. The researchers also developed a number of applications of this DNAzyme. They introduced a new paradigm for the sensitive analysis of DNA by using a DNA-based machine that consists of a DNA template, polymerization/nicking enzymes, and strand displacement of the peroxidase-mimicking DNAzyme [58]. The isothermal and rapid (approximately 90 min) analysis of the target DNA, together with the quantitative detection of the analyte, demonstrate the appealing bioanalytical features of the method. They also employed a circular DNA as a template for the amplified detection of M13 phage ssDNA by a rolling circle amplification (RCA) process that synthesizes peroxidase-mimicking DNAzyme chains, thus enabling the colorimetric or chemiluminescent detection of the analyte [59]. This method had two amplification steps: the RCA synthesis of numerous DNAzyme units, as a result of a single recognition (hybridization) event, and the catalytic activities of the synthesized labels that lead to the colorimetric or chemiluminescent detection of DNA. A detection limit of 1×10^{-14} M was reported for this method.

Another kind of DNAzyme is a divalent metal ion cofactor-specific. Consequently, the development of DNAzyme-based sensors with that characteristic has begun to attract much attention to detect a variety of environmentally relevant metal ions, such as Cu^{2+} [60–63], Zn^{2+} [64–66], Pb^{2+} [67–70], Hg^{2+} [23, 71], UO_2^{2+}

[72] and Ca^{2+} [73]. Heavy metals are ubiquitous environmental contaminants, and their high toxicity makes their presence undesirable. Generally, metal ions are classified into two categories: essential and nonessential. Nonessential heavy metals like lead (Pb) and mercury (Hg) can cause a number of adverse health effects even at low-level of exposure. Essential metals such as copper (Cu) and zinc (Zn) are required to support biological activities. However, even these essential metals are toxic in excess [74]. A large number of heavy metals can cause disease or even death, constituting a serious threat to human health. Cu^{2+} and Pb^{2+} also pose serious environmental problems and are potentially toxic for all living organisms like other heavy metal ions. Due to their toxicity, the U.S. EPA has set the MAL-of copper and lead in drinking water as 20 μM (1.3 ppm) and 72 nM (15 ppb), respectively. Therefore, accurate determination of these metals at the trace level in environmental and biological samples has become increasingly important. Correspondingly, a number of analysis methods for the detection of heavy metal ions have been developed over the last few years, including not only atomic absorption spectroscopy [75], voltammetric detection [76], and inductively coupled plasma mass spectrometry [77] as classical detection methods, but also some other strategies such as sensors [78], surface plasmon resonance (SPR) [79], peptide [80], and peptide coated quantum dots [81]. Lu's group developed fluorescent sensors using DNAzyme for the detection of a number of metal ions. These studies demonstrated high sensitivity and selectivity. For example, they reported nanoparticle-based colorimetric sensors using DNAzyme-catalyzed ligation reaction for Cu^{2+} with a detection limit of ~ 5 μM [63]. Furthermore, they developed fluorescent metal sensors using a Cu^{2+}-dependent DNA-cleaving DNAzyme with a detection limit of 35 nM (2.3 ppb) [82].

In this section, we present a novel highly sensitive and selective FPA for detecting Cu^{2+} and Pb^{2+} ion based on metal ion-dependent DNAzyme [83]. The detection sensitivity of Cu^{2+} and Pb^{2+} can be significantly improved to ~ 1 nM (Ca. 65 and 200 ppt) by using a "nanoparticle enhancement" approach. Figure 2.4 outlines the principle to use fluorescence anisotropy assay enhanced by AuNPs for the Cu^{2+} ion or Pb^{2+} ion detection. From Perrin equation, the fluorescence anisotropy (r) is sensitive to changes in the rotational motion of fluorescently labeled molecules. The r value of a fluorophore is proportional to its rotational relaxation time, which in turn depends upon its molecular volume (molecular weight). On the basis of Cu^{2+}-dependent DNA-cleaving DNAzyme, we first designed and prepared two DNA strands: DNAzyme named as Cu-Enz and its substrate DNA strand named as Cu-Sub. Cu-Sub DNA labeled with 6-fluorescein-CE phosphoramidite (FAM) at the $3'$ end and thiol group at the $5'$ end (Cu-Sub, $5'$ SH-T$_{20}$AGCTTCTTTCTAATACGGCTTACC-FAM3$'$) was immobilized onto the surface of gold nanoparticle (13 nm) via a thiol-Au interaction. The Cu-Enz and Cu-Subs on AuNPs form the complexes through two base-pairing regions; the $5'$ portion of the DNAzyme binds the substrate via Watson–Crick base pairs and the $3'$-region through formation of a DNA triplex (Fig. 2.4d). The FAM fluorophore in the "big" AuNPs-mediated Cu-Sub and Cu-Enz complexes displays the high anisotropy (Fig. 2.4a, b). In the presence of Cu^{2+}, the Cu-Enz as a recognition

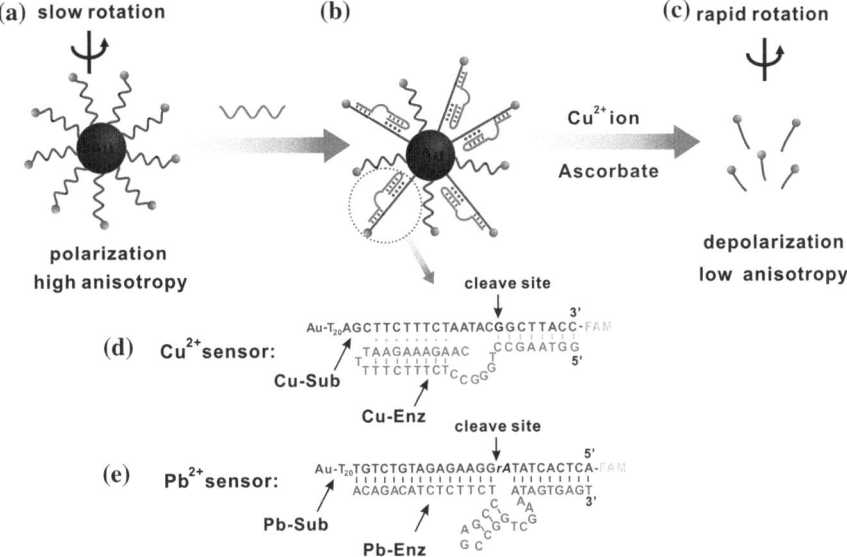

Fig. 2.4 Schematic illustration of the strategy of Cu^{2+} ion or Pb^{2+} ion detection using Cu^{2+}-dependent DNA-cleaving DNAzyme or Pb^{2+}-dependent DNA-cleaving DNAzyme fluorescence anisotropy via gold nanoparticles (AuNPs) enhancement. The point of scission is indicated by black arrow. Structural model and sequences for bimolecular DNA complexes of DNAzyme and its substrate for Cu^{2+} ion and Pb^{2+} ion are presented. The point of scission is indicated by black arrow. Reproduced from Ref. [83], with the permission of Elsevier Inc

element specifically cleaves AuNPs-mediated Cu-Sub at the guanine base (marked in black). The AuNPs-mediated Cu-Sub and Cu-Enz complexes are separated due to lacking thermal stability. The FAM-labeled cleaved products are released in the solution with rapid rotation and depolarize the light (Fig. 2.4c). Thus the anisotropy of the system will significantly decrease correspondingly. In the absence of Cu^{2+}, the AuNPs-mediated Cu-Sub and Cu-Enz associates would have stable high anisotropy. Therefore via the free fluorescent cleavage products, this cleave event is translated into a measurable fluorescence anisotropy decrease proportional to Cu^{2+} concentration. This design can also be easily expanded to Pb^{2+} ion detection, just displace the Cu-Sub and Cu-Enz with Pb-Sub and Pb-Enz. The structural model and sequences for bimolecular DNA complexes of Pb^{2+}-dependent DNA-cleaving DNAzyme and its substrate are also presented in Fig. 2.4e.

In order to verify the specificity of DNA-cleaving DNAzyme relied on the Cu^{2+} ion, we performed the control experiments to investigate if the DNA molecules (Cu-Sub) undergo cleavage in the absence of corresponding DNAzyme (Cu-Enz) and if the Cu^{2+} ion is used as the only cofactor. Three group experiments were carried out as follow: titrating Cu^{2+} ion, Hg^{2+} ion, or Pb^{2+} ion to the Cu-Sub functionalized AuNPs solution (Cu-Enz added), titrating Cu^{2+} ion, Hg^{2+} ion or Pb^{2+} ion to the Cu-Sub functionalized AuNPs solution (no Cu-Enz added), and titrating Cu^{2+} ion, Hg^{2+} ion, or Pb^{2+} ion to Cu-Sub functionalized AuNPs solution with fully complementary DNA (Cu-Sub-Comp, 5′GGTAAGCCGTATT AGA-AAGAAGCT 3′). There was little anisotropy change when titrating Cu^{2+}, Hg^{2+} or Pb^{2+} ion to the Cu-Sub functionalized AuNPs (Cu-Enz added) solution. The results were the same when the tested metal ions were added to the solution containing Cu-Sub functionalized AuNPs and theircomplementary DNA. In contrast, when Cu-Enz coexisted with Cu-Sub, there would be a significant anisotropy change as Cu^{2+} ion was added.

As a proof-of-concept experiment, a novel and practical method relying on metal ion-dependent DNAzyme-based fluorescence anisotropy via AuNPs enhancement was presented for detecting Cu^{2+} and Pb^{2+} ions in aqueous media at room temperature. The FAM-labeled DNA substrates in the "big" AuNPs-mediated complexes were cleaved and released. The AuNPs-enhancement function results in a significant decrease. This effect can be used for the Cu^{2+} and Pb^{2+} detection as low as 1 nM. This method opens up a new possibility of rapid and easy detection of toxic metal ions in environmental samples. Indeed, for a practical application purpose, preliminary experiments were performed on the real river water samples and the spiked river samples. The results revealed good recoveries. This method shows distinct advantages over conventional methods in terms of its potential sensitivity, specificity, and ability for rapid response.

2.4 Peptide-Functionalized Spherical Polyelectrolyte Nanobrushes for Real-Time Sensing of Protease Activity

Proteases play vital roles in physiological processes from simple protein catabo-lism to highly regulated cascades, such as blood coagulation, cell regulation, and signaling and immune defense, as well as pathological processes, including inflammation and tumor growth [84–86]. Consequently, interest in the activity of proteases, as well as their inhibition mechanism, is increasingly directed toward biochemical research, diagnosis of protease-related diseases, and the development of potential therapeutic drugs. Over the past decade, many methods for detecting protease activity have been reported, such as high-performance liquid chromatography (HPLC) [87], polyacrylamide gel electrophoresis [88], capillary electrophoresis [89], and FRET-based methods (fluorescence resonance energy transfer) [90, 91], but most of them are time-consuming and impractical for a real-time, multiplex, or high-throughput format. Therefore, there is a high demand for a simple, rapid, sensitive, and high-throughput routine assay to detect protease activity. Recently, some inorganic nanomaterials, such as quantum dots and gold nanoparticles, have been widely explored for use in protease assays [92–94].

In this section, we demonstrate the peptide-functionalized spherical polyelec-trolyte brushes (SPB), as a new type of biocompatible colloidal nanostructure with good water-solubility, can be exploited to generate a highly sensitive FPA for protease activity sensing [95]. SPB consists of a solid polystyrene core with a diameter of 50–100 nm onto which linear polyelectrolyte poly(acrylic acid) chains

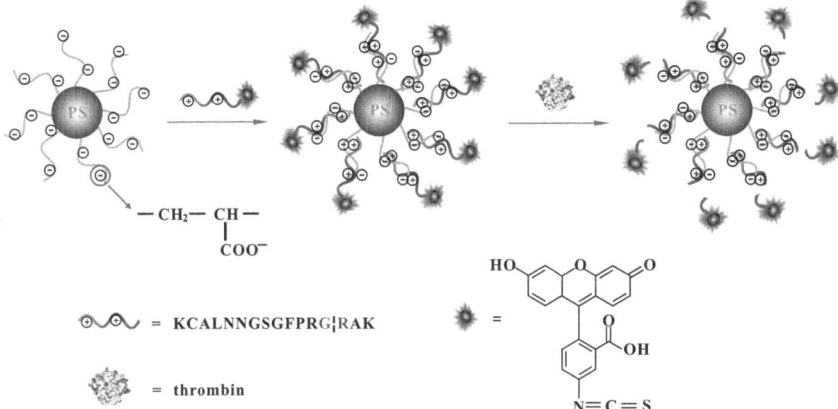

Fig. 2.5 Schematic illustration of the strategy using Peptide-SPB composite probes as substrates to monitor the proteolytic activity of thrombin. The colloidal particles consist of a solid polystyrene (*PS*) core onto which cationic polyelectrolyte poly(acrylic acid) (*PAA*) chains are grafted. The peptide was labeled with fluorescein isothiocyanate (*FITC*). The cleavage point at Arg-Gly bonds are indicated by dashed line. The chemical structural formula of FITC and its coupling site with peptide is provided at the bottom right. FITC is the base fluorescein molecule functionalized with an isothiocyanate reactive group (–N = C = S) on the bottom ring of the structure, which is reactive toward amine group of the lysine. Reproduced from Ref. [95], with the permission of Wiley John and Sons

(PAA) are grafted [96]. Due to its excellent dispersibility and good retention, PAA has been shown to be an attractive carrier for drug delivery systems, biosensors, immunoassays, and biocatalysts. Ballauff's group has done a systematic work of nanobrushes (SPB) in the biological applications [97–100], especially in the protein immobilization onto the SPB. Fig. 2.5 illustrates the principle of the SPB-based FPA proposed in this work. Briefly, fluorophore-labeled peptide substrates are rapidly assembled onto the thin polyelectrolyte brushes to form Peptide-SPB composite probes, which, in turn, are capable of displaying a high FP value. The *FP* value of a fluorophore is sensitive to changes in the rotational motion of fluorescently labeled molecule. Fluorophore-labeled peptides confined in SPB nanobrushes will have bigger anisotropy values due to their slow rotation. In the presence of protease, the Peptide-SPB composite probes are recognized and specifically cleaved to release the fluorophore-labeled peptide segments, resulting in depolarization of the system. This SPB-peptide hydrolysis event is then translated into measurable *FP* value decrease by free cleaved dye-labeled peptide segments, which results in the sensing of protease activity.

The most important feature of polyelectrolyte (PE) brushes is the strong confinement of counterions within the brush and their distribution along the polymer backbone. We investigated the adsorption of different biomolecules (nucleic acids, peptides, and protein) from the solution onto the brush shell chains of SPB. As expected, more than 90% of the positively charged peptides were adsorbed onto the

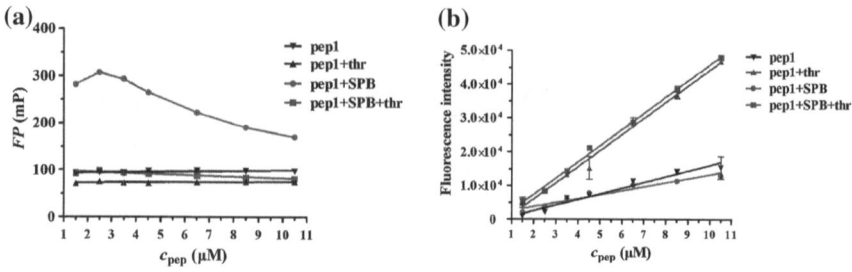

Fig. 2.6 Investigation of binding stoichiometry and fluorescence intensity (*FI*) for Peptide-SPB composite probes and free peptide1. **a** The change of *FP* value obtained from the increased peptide 1 in the presence or absence of SPB (10 nM) and in the presence or absence of thrombin (2.5 units). **b** The change of fluorescence intensity (*FI*) obtained from the increased peptide1 in the presence or absence of SPB (10 nM) and in the presence or absence of thrombin (2.5 units). The data were from the time point of 4 h after the proteolytic reaction. Reproduced from Ref. [95], with the permission of Wiley

SPB particles in a quick equilibration within 10 min, which also showed the excellent stability of Peptide-SPB composite probes against coagulation and leaching out of the peptides. Next, we investigated the binding stoichiometry of Peptide-SPB composite probes using FITC-labeled peptide 1 (Lys-Cys-Ala-Leu-Asn–Asn-Gly-Ser-Gly-dPhe-Pro-Arg-Gly-Arg-Ala-Lys(FITC)-OH), which was found to be one of the best thrombin substrates. As shown in Fig. 2.6, the binding stoichiometry of Peptide-SPB composites probes was estimated by evaluating the *FP* value responses of the Peptide-SPB composite probes to different concentration ratios of SPB and peptide 1. The number of peptides per SPB nanoparticle was estimated to be 250. In addition, it is worth noting that we observed the pronounced increasing FI of the mixture caused by more and smaller fluorogenic peptide segments released as the proteolytic reaction proceeded.

The method was tested with thrombin, a serine protease playing a pivotal role in hemostasis and blood clotting by selectively cleaving Arg–Gly bonds in fibrinogen to form fibrin and platelet activation [101]. The FITC-labeled peptide 1 (Lys-Cys-Ala-Leu-Asn–Asn-Gly-Ser-Gly-dPhe-Pro-Arg-Gly-Arg-Ala-Lys(FITC)-OH), of which the core peptide chain dPhe-Pro-Arg-Gly was designed according to the reported material [102], was determined to be one of the best thrombin substrates. Therefore, the prepared peptide 1-functionalized SPB (Peptide-SPB composite probes), corresponding to an initial *FP* value of ∼310 mP, were utilized for thrombin activity detection. Addition of thrombin (2.5 units) to the Peptide-SPB composite probes solutions with different concentrations resulted in time-dependent decreases in FP (Fig. 2.7a). Specifically, the peptides of Peptide-SPB composite probes were selectively cleaved at the Arg-Gly bonds gradually, and then the cleaved FITC-labeled peptide was released from the SPB to the solution with faster rotation, resulting in depolarizing the system and lowering the *FP* value. From these data, the kinetic parameters for hydrolysis of Peptide-SPB composite probes by thrombin were determined (Fig. 2.7b). The V_{max} was calculated to be

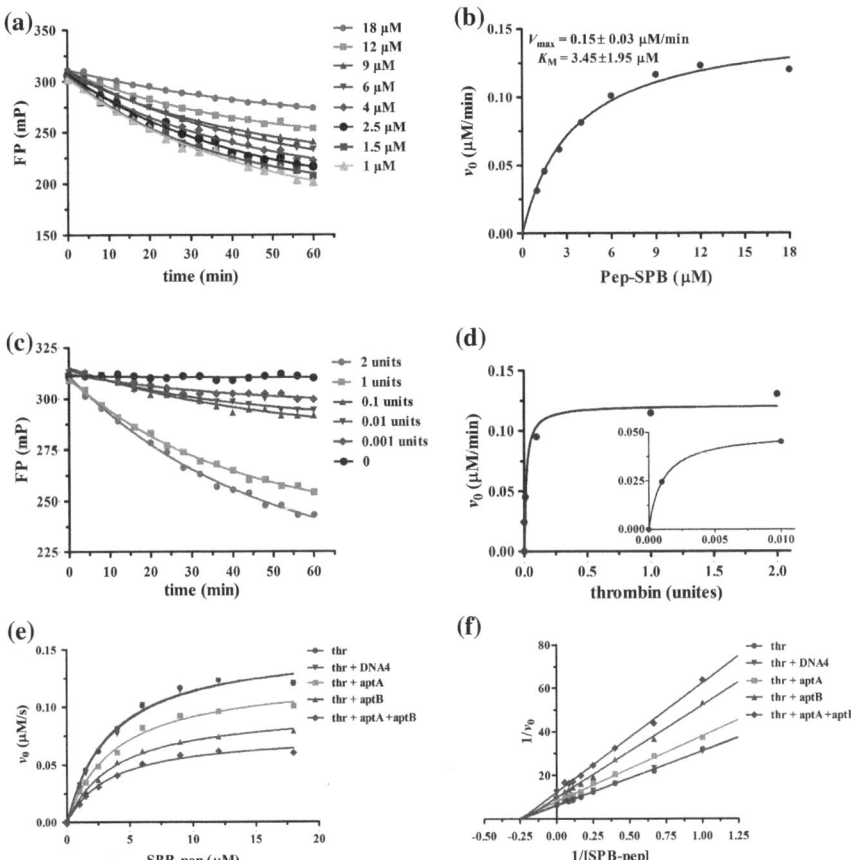

Fig. 2.7 a Progress curves at different concentrations of Peptide-SPB composite probes (1–18 μM, corresponding to peptide concentrations). Thrombin of 2.5 units was used. **b** Plot of the initial velocities for an increasing concentration of Peptide-SPB composite probes at a constant amount of thrombin (2.5 units), and the data were fit by nonlinear least-squares to the Michaelis–Menten equation. **c** Progress curves at different concentrations of thrombin. **d** Plots of the initial velocities for the studied concentrations of thrombin. Peptide-SPB composite probes of 6 μM were used. **e** Plots of the initial velocities against Peptide-SPB composite probes in the absence and presence of aptA, aptB, or DNA 4, and the data were fit by nonlinear least-squares to the Michaelis–Menten equation. **f** Lineweaver–Burk (L-B) double reciprocal plot of the same data for determination of K_i values. Reproduced from Ref. [95], with the permission of Wiley John and Sons

0.15 ± 0.03 μM/min, and the k_{cat} was 0.30 min^{-1}, which were comparable to the literature values of 0.2 ± 0.01 μM/min and 0.20 min^{-1} [94]. The K_M value was 3.45 ± 1.95 μM, which was 7.5-fold greater than the reported value of 0.46 ± 0.12 μM [94]. The high K_M value may have resulted from the phenomenon in which some peptides of the innermost SPB brushes are inaccessible for thrombin. The SPB-based FPA is also suitable for quantitative determination of the amount of thrombin present in solution. Fig. 2.7c illustrates the real-time change in *FP* value with different concentrations of thrombin. The assay allowed for the detection of thrombin activity at concentrations as low as 1.0×10^{-3} units ($1.85 \sim 5.55$ nM) based on three times the standard deviation (3σ). At a thrombin concentration of 1.0×10^{-3} units ($1.85 \sim 5.55$ nM), the rate of the hydrolysis of peptide was 0.0245 μM/min (Fig. 2.7d). This detection limit was better than the recently reported gold nanoparticle methods of 20 nM [5] and 5 nM [92], as well as electrochemical sensors of 1 nM [103] and 0.5 nM [104], considering the low enzyme activity used in this work (50–150 NIH units/mg protein). The substantial improvement of our method in the sensitivity is mainly attributed to the slower rotation of fluorescence unit when dye-labeled peptides were confined in the nanobrushes shell of SPB.

Interest in the screening of thrombin inhibitors is rapidly increasing for anticoagulant drug development. Therefore, the inhibition of thrombin activity by the two known anti-thrombin aptamers was investigated using the SPB-based FPA. One is 15-mer ssDNA (aptA: 5′GGTTGGTGTGGTTGG3′), which forms an intramolecular quadruplex structure and primarily binds to thrombin at the fibrinogen-recognition exosite [105, 106]. The other one is 29-mer ssDNA (aptB: 5′AGTCCGTGGTAGGG CAGGTTGGGGTGACT3′) binding at the heparin-binding exosite with a higher affinity [107]. Fig. 2.7e shows representative plots of velocity derived from monitoring changes in the FP value for increasing concentrations of Peptide-SPB composite probes exposed to the thrombin (2.5 units) and the aptamers (10 μM). The inhibition constant (K_i) values for aptA and aptB were estimated to be 40 ± 15 and 20 ± 14 μM, respectively (Fig. 2.7f). When the two aptamers were exposed to the thrombin simultaneously, the K_i value was reduced to 11 ± 4 μM, which was much smaller than any one of the individual aptamers alone. This result clearly indicated that the two aptamers have different binding sites on the thrombin. At the same time, we observed that the random nucleic acid of DNA 4 had no effect on thrombin, while aptB displayed a stronger inhibitory function of thrombin than aptA, which was in good agreement with the reported result [107].

Following the successful demonstration using the SPB-based FPA to study the inhibition of two anti-thrombin aptamers, we further investigated the applicability of the method for the high-throughput screening of a library of inhibitors. To achieve this, four known thrombin inhibitors, including α-iodoacetamide, sodium fluoride, and two aptamers, together with a set of 103 organic compounds (terpenoids, neonicotinoids, benzothiazinoids), were used to screen the potential inhibitors of thrombin. The experiment was conveniently carried out on a 384-well plate with all samples in duplicate. Because the organic compounds are insoluble in water, all compounds were dissolved in DMSO, which is commonly used as a co-solvent. First, we assessed the influence of DMSO concentration on the FP signal in this work. The experimental results demonstrated that DMSO concentrations up to 2% had no effect on FP measurement and were well tolerated by thrombin and trypsinase (data not shown). In the screening experiment, we selected the negative-control reactions without added compounds and positive-control reactions with a known thrombin inhibitor to set low and high boundaries for FP signals, respectively. In this work, a "hit" was defined as compounds that reduced the FP signal by 50% relative to control reactions. As expected, the

Fig. 2.8 Screening for thrombin inhibitors. A screening of 107 compounds identified 4 hits (*1* aptB, *2* aptA, *3* α-iodoacetamide, *4* sodium fluoride) that reduced the *FP* signal by 50% relative to control reactions in the absence of added compound. Reproduced from Ref. [95], with the permission of Wiley John and Sons

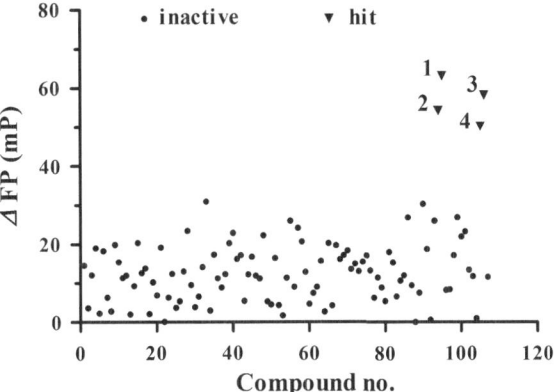

α-iodoacetamide (51.8%), sodium fluoride (60.1%), aptA (55.9%), and aptB (65.2%) all showed obvious inhibitory effects on the thrombin, while none of the organic compounds exhibited a significant inhibitory effect on thrombin (Fig. 2.8). These results, in turn, show that the method is suitable for high-throughput applications in the drug discovery of specific protease inhibitors.

In summary, we developed a simple and sensitive protease assay that utilizes the self-assembled Peptide-SPB probe to enhance the FP signal, which enables the real-time detection of protease activity. This novel SPB-peptide probe design offers many advantages, including simplicity and rapidity of preparation and manipulation compared to methods employing specific synthetic strategies. The SPB-based FPA demonstrates superior sensitivity. The LOD of thrombin activity could be significantly improved to 1.0×10^{-3} units (1.85 ~ 5.55 nM), indicating that our system is one of the most sensitive protease detection systems. Moreover, our Peptide-SPB probe design can be extended to develop a variety of probes by simply changing the peptide sequence for measuring the activity of other proteolytic enzymes and kinases. The assay can be carried out in 96- or 384-well plates, making it suitable for routine high-throughput applications in drug development and biotechnology. Overall, our results demonstrate that the colloidal spherical PE brush, as a new class of nanostructures, can provide an excellent platform for the development of rapid and sensitive assays for sensing protease activity and high-throughput screening of inhibitors.

References

1. Weber G (1952) Polarization of the fluorescence of macromolecules. II. Fluorescent conjugates of ovalbumin and bovine serum albumin. Biochem J 51(2):155–167
2. Jameson DM, Ross JA (2010) Fluorescence polarization/anisotropy in diagnostics and imaging. Chem Rev 110(5):2685–2708. doi:10.1021/cr900267p
3. Perrin F (1926) Polarization of light of fluorescence, average life of molecules in the excited state. J Phys Radium 7(12):390–401. doi:10.1051/jphysrad:01926007012039000

4. Weizmann Y, Patolsky F, Willner I (2001) Amplified detection of DNA and analysis of single-base mismatches by the catalyzed deposition of gold on Au-nanoparticles. Analyst 126(9):1502–1504. doi:10.1039/B106613G

5. Pavlov V, Xiao Y, Shlyahovsky B, Willner I (2004) Aptamer-functionalized Au nanoparticles for the amplified optical detection of thrombin. J Am Chem Soc 126(38):11768–11769. doi:10.1021/ja046970u

6. Yan Y-M, Tel-Vered R, Yehezkeli O, Cheglakov Z, Willner I (2008) Biocatalytic growth of Au nanoparticles immobilized on glucose oxidase enhances the ferrocene-mediated bioelectrocatalytic oxidation of glucose. Adv Mater 20(12):2365–2370. doi:10.1002/adma. 200703128

7. Dong X, Lau CM, Lohani A, Mhaisalkar SG, Kasim J, Shen Z, Ho X, Rogers JA, Li L-J (2008) Electrical detection of femtomolar DNA via gold-nanoparticle enhancement in carbon-nanotube-network field-effect transistors. Adv Mater 20(12):2389–2393. doi:10. 1002/adma.200702798

8. Zhang J, Song S, Zhang L, Wang L, Wu H, Pan D, Fan C (2006) Sequence-specific detection of femtomolar DNA via a chronocoulometric DNA sensor (CDS): Effects of nanoparticle-mediated amplification and nanoscale control of DNA assembly at electrodes. J Am Chem Soc 128(26):8575–8580. doi:10.1021/ja061521a

9. Tokarev I, Tokareva I, Minko S (2008) Gold-nanoparticle-enhanced plasmonic effects in a responsive polymer gel. Adv Mater 20(14):2730–2734. doi:10.1002/adma.200702885

10. Harris HH, Pickering IJ, George GN (2003) The chemical form of mercury in fish. Science 301(5637):1203. doi:10.1126/science.1085941

11. Clarkson TW, Magos L, Myers GJ (2003) The toxicology of mercury—current exposures and clinical manifestations. New Engl J Med 349(18):1731–1737. doi:10.1056/NEJMra022471

12. Morel FMM, Kraepiel AML, Amyot M (1998) The chemical cycle and bioaccumulation of mercury. Annu Rev Ecol Syst 29(1):543–566. doi:10.1146/annurev.ecolsys.29.1.543

13. Yang Y-K, Yook K-J, Tae J (2005) A rhodamine-based fluorescent and colorimetric chemodosimeter for the rapid detection of Hg^{2+} ions in aqueous media. J Am Chem Soc 127(48):16760–16761. doi:10.1021/ja054855t

14. Nazeeruddin MK, DiCenso D, Humphry-Baker R, Grätzel M (2006) Highly selective and reversible optical, colorimetric, and electrochemical detection of mercury(II) by amphiphilic ruthenium complexes anchored onto mesoporous oxide films. Adv Funct Mater 16(2):189–194. doi:10.1002/adfm.200500309

15. Nolan EM, Lippard SJ (2007) Turn-on and ratiometric mercury sensing in water with a red-emitting probe. J Am Chem Soc 129(18):5910–5918. doi:10.1021/ja068879r

16. Avirah RR, Jyothish K, Ramaiah D (2006) Dual-mode semisquaraine-based sensor for selective detection of Hg^{2+} in a micellar medium. Org Lett 9(1):121–124. doi:10.1021/ol062691v

17. Lee MH, Cho B-K, Yoon J, Kim JS (2007) Selectively chemodosimetric detection of Hg(II) in aqueous media. Org Lett 9(22):4515–4518. doi:10.1021/ol7020115

18. Othman AB, Lee JW, Wu J-S, Kim JS, Abidi R, Thuéry P, Strub JM, Van Dorsselaer A, Vicens J (2007) Calix[4]arene-based, Hg^{2+}-induced intramolecular fluorescence resonance energy transfer chemosensor. J Org Chem 72(20):7634–7640. doi:10.1021/jo071226o

19. Coskun A, Akkaya EU (2006) Signal ratio amplification via modulation of resonance energy transfer: Proof of principle in an emission ratiometric Hg(II) sensor. J Am Chem Soc 128(45):14474–14475. doi:10.1021/ja066144g

20. Wang J, Qian X (2006) A series of polyamide receptor based PET fluorescent sensor molecules: Positively cooperative Hg^{2+} ion binding with high sensitivity. Org Lett 8(17):3721–3724. doi:10.1021/ol061297u

21. Yoon S, Miller EW, He Q, Do PH, Chang CJ (2007) A bright and specific fluorescent sensor for mercury in water, cells, and tissue. Angew Chem Int Ed 46(35):6658–6661. doi:10.1002/ anie.200701785

22. Kim I-B, Bunz UHF (2006) Modulating the sensory response of a conjugated polymer by proteins: An agglutination assay for mercury ions in water. J Am Chem Soc 128(9): 2818–2819. doi:10.1021/ja058431a

23. Liu J, Lu Y (2007) Rational design of "turn-on" allosteric DNAzyme catalytic beacons for aqueous mercury ions with ultrahigh sensitivity and selectivity. Angew Chem Int Ed 46(40):7587–7590. doi:10.1002/anie.200702006

24. Darbha GK, Singh AK, Rai US, Yu E, Yu H, Chandra Ray P (2008) Selective detection of mercury (II) ion using nonlinear optical properties of gold nanoparticles. J Am Chem Soc 130(25):8038–8043. doi:10.1021/ja801412b

25. Li H, Zhang Y, Wang X, Xiong D, Bai Y (2007) Calixarene capped quantum dots as luminescent probes for Hg^{2+} ions. Mater Lett 61(7):1474–1477. doi:10.1016/j.matlet.2006.07.064

26. Chen P, He C (2003) A general strategy to convert the MerR family proteins into highly sensitive and selective fluorescent biosensors for metal ions. J Am Chem Soc 126(3):728–729. doi:10.1021/ja0383975

27. Hakkila K, Green T, Leskinen P, Ivask A, Marks R, Virta M (2004) Detection of bioavailable heavy metals in EILATox-Oregon samples using whole-cell luminescent bacterial sensors in suspension or immobilized onto fibre-optic tips. J Appl Toxicol 24(5):333–342. doi:10.1002/jat.1020

28. Ono A, Togashi H (2004) Highly selective oligonucleotide-based sensor for mercury(II) in aqueous solutions. Angew Chem Int Ed 43(33):4300–4302. doi:10.1002/anie.200454172

29. Lee J-S, Han MS, Mirkin CA (2007) Colorimetric detection of mercuric ion (Hg^{2+}) in aqueous media using DNA-functionalized gold nanoparticles. Angew Chem Int Ed 46(22):4093–4096. doi:10.1002/anie.200700269

30. Xue X, Wang F, Liu X (2008) One-step, room temperature, colorimetric detection of mercury (Hg^{2+}) using DNA/nanoparticle conjugates. J Am Chem Soc 130(11):3244–3245. doi:10.1021/ja076716c

31. Li D, Wieckowska A, Willner I (2008) Optical analysis of Hg^{2+} ions by oligonucleotide-gold-nanoparticle hybrids and DNA-based machines. Angew Chem Int Ed 47(21):3927–3931. doi:10.1002/anie.200705991

32. Ye BC, Yin BC (2008) Highly sensitive detection of mercury(II) ions by fluorescence polarization enhanced by gold nanoparticles. Angew Chem Int Ed 47(44):8386–8389. doi:10.1002/anie.200803069

33. Imahori H, Kashiwagi Y, Endo Y, Hanada T, Nishimura Y, Yamazaki I, Araki Y, Ito O, Fukuzumi S (2004) Structure and photophysical properties of porphyrin-modified metal nanoclusters with different chain lengths. Langmuir 20(1):73–81. doi:10.1021/la035435p

34. Guerouï Z, Libchaber A (2004) Single-molecule measurements of gold-quenched quantum dots. Phys Rev Lett 93(16):166108. doi:10.1103/PhysRevLett.93.166108

35. Dulkeith E, Ringler M, Klar TA, Feldmann J, Munoz Javier A, Parak WJ (2005) Gold nanoparticles quench fluorescence by phase induced radiative rate suppression. Nano Lett 5(4):585–589. doi:10.1021/nl0480969

36. Dubertret B, Calame M, Libchaber AJ (2001) Single-mismatch detection using gold-quenched fluorescent oligonucleotides. Nat Biotechnol 19(4):365–370. doi:10.1038/86762

37. Maxwell DJ, Taylor JR, Nie S (2002) Self-assembled nanoparticle probes for recognition and detection of biomolecules. J Am Chem Soc 124(32):9606–9612. doi:10.1021/ja025814p

38. Du H, Disney MD, Miller BL, Krauss TD (2003) Hybridization-based unquenching of DNA hairpins on au surfaces: Prototypical "molecular beacon" biosensors. J Am Chem Soc 125(14):4012–4013. doi:10.1021/ja0290781

39. Li H, Rothberg LJ (2004) DNA sequence detection using selective fluorescence quenching of tagged oligonucleotide probes by gold nanoparticles. Anal Chem 76(18):5414–5417. doi:10.1021/ac049173n

40. Schneider G, Decher G, Nerambourg N, Praho R, Werts MH, Blanchard-Desce M (2006) Distance-dependent fluorescence quenching on gold nanoparticles ensheathed with layer-by-layer assembled polyelectrolytes. Nano Lett 6(3):530–536. doi:10.1021/nl052441s

41. Sokolov K, Chumanov G, Cotton TM (1998) Enhancement of molecular fluorescence near the surface of colloidal metal films. Anal Chem 70(18):3898–3905. doi:10.1021/ac9712310

42. Kulakovich O, Strekal N, Yaroshevich A, Maskevich S, Gaponenko S, Nabiev I, Woggon U, Artemyev M (2002) Enhanced luminescence of CdSe quantum dots on gold colloids. Nano Lett 2(12):1449–1452. doi:10.1021/nl025819k

43. Malicka J, Gryczynski I, Fang J, Kusba J, Lakowicz JR (2003) Increased resonance energy transfer between fluorophores bound to DNA in proximity to metallic silver particles. Anal Biochem 315(2):160–169. doi:10.1016/S0003-2697(02)00710-8

44. Matveeva EG, Gryczynski I, Barnett A, Leonenko Z, Lakowicz JR, Gryczynski Z (2007) Metal particle-enhanced fluorescent immunoassays on metal mirrors. Anal Biochem 363(2):239–245. doi:10.1016/j.ab.2007.01.030

45. Gerber S, Reil F, Hohenester U, Schlagenhaufen T, Krenn JR, Leitner A (2007) Tailoring light emission properties of fluorophores by coupling to resonance-tuned metallic nanostructures. Phys Rev B 75(7):073404. doi:10.1103/PhysRevB.75.073404

46. Kruger K, Grabowski PJ, Zaug AJ, Sands J, Gottschling DE, Cech TR (1982) Self-splicing RNA: Autoexcision and autocyclization of the ribosomal RNA intervening sequence of tetrahymena. Cell 31(1):147–157. doi:10.1016/0092-8674(82)90414-7

47. Guerrier-Takada C, Gardiner K, Marsh T, Pace N, Altman S (1983) The RNA moiety of ribonuclease P is the catalytic subunit of the enzyme. Cell 35 (3, Part 2):849-857. doi:10.1016/0092-8674(83)90117-4

48. Symons RH (1992) Small catalytic RNAs. Annu Rev Biochem 61:641–671. doi:10.1146/annurev.bi.61.070192.003233

49. Thomas RC (1983) RNA splicing: three themes with variations. Cell 34(3):713–716. doi:10.1016/0092-8674(83)90527-5

50. Ban N (2000) The complete atomic structure of the large ribosomal subunit at 2.4 angstrom resolution. Science 289 (5481):905. doi:10.1126/science.289.5481.905

51. Breaker RR, Joyce GF (1994) A DNA enzyme that cleaves RNA. Chem Biol 1(4):223–229. doi:10.1016/1074-5521(94)90014-0

52. Wilson DS, Szostak JW (1999) In vitro selection of functional nucleic acids. Annu Rev Biochem 68(1):611–647. doi:10.1146/annurev.biochem.68.1.611

53. Breaker RR (1997) DNA enzymes. Nat Biotech 15(5):427–431

54. Ronald RB (1997) DNA aptamers and DNA enzymes. Curr Opin Chem Biol 1(1):26–31. doi:10.1016/s1367-5931(97)80105-6

55. Lu Y (2002) New transition-metal-dependent DNAzymes as efficient endonucleases and as selective metal biosensors. Chem-Eur J 8(20):4588–4596. doi:10.1002/1521-3765(20021018)8:20<4588:aid-chem4588>3.0.co;2-q

56. Liu J, Cao Z, Lu Y (2009) Functional nucleic acid sensors. Chem Rev 109(5):1948–1998. doi:10.1021/cr030183i

57. Niazov T, Pavlov V, Xiao Y, Gill R, Willner I (2004) DNAzyme-functionalized Au nanoparticles for the amplified detection of DNA or telomerase activity. Nano Lett 4(9):1683–1687. doi:10.1021/nl0491428

58. Weizmann Y, Beissenhirtz MK, Cheglakov Z, Nowarski R, Kotler M, Willner I (2006) A virus spotlighted by an autonomous DNA machine. Angew Chem Int Ed 45(44): 7384–7388. doi:10.1002/anie.200602754

59. Cheglakov Z, Weizmann Y, Basnar B, Willner I (2007) Diagnosing viruses by the rolling circle amplified synthesis of DNAzymes. Org Biomol Chem 5(2):223–225

60. Carmi N, Balkhi SR, Breaker RR (1998) Cleaving DNA with DNA. Proc Natl Acad Sci U S A 95(5):2233–2237

61. Carmi N, Breaker RR (2001) Characterization of a DNA-Cleaving deoxyribozyme. Bioorg Med Chem 9(10):2589–2600. doi:10.1016/s0968-0896(01)00035-9

62. Carmi N, Shultz LA, Breaker RR (1996) In vitro selection of self-cleaving DNAs. Chem Biol 3(12):1039–1046. doi:10.1016/S1074-5521(96)90170-2

63. Liu J, Lu Y (2007) Colorimetric Cu^{2+} detection with a ligation DNAzyme and nanoparticles. Chem Commun 46:4872–4874. doi:10.1039/B712421J

64. Li J, Zheng W, Kwon AH, Lu Y (2000) In vitro selection and characterization of a highly efficient Zn(II)-dependent RNA-cleaving deoxyribozyme. Nucleic Acids Res 28(2):481–488. doi:10.1093/nar/28.2.481

65. Santoro SW, Joyce GF, Sakthivel K, Gramatikova S, Barbas CF (2000) RNA cleavage by a DNA enzyme with extended chemical functionality. J Am Chem Soc 122(11):2433–2439. doi:10.1021/ja993688s

66. Kim H-K, Liu J, Li J, Nagraj N, Li M, Pavot CMB, Lu Y (2007) Metal-dependent global folding and activity of the 8-17 DNAzyme studied by fluorescence resonance energy transfer. J Am Chem Soc 129(21):6896–6902. doi:10.1021/ja0712625

67. Liu J, Lu Y (2003) A colorimetric lead biosensor using DNAzyme-directed assembly of gold nanoparticles. J Am Chem Soc 125(22):6642–6643. doi:10.1021/ja034775u

68. Yim T-J, Liu J, Lu Y, Kane RS, Dordick JS (2005) Highly active and stable DNAzyme-carbon nanotube hybrids. J Am Chem Soc 127(35):12200–12201. doi:10.1021/ja0541581

69. Xiao Y, Rowe AA, Plaxco KW (2006) Electrochemical detection of parts-per-billion lead via an electrode-bound DNAzyme assembly. J Am Chem Soc 129(2):262–263. doi:10.1021/ja067278x

70. Shen L, Chen Z, Li Y, He S, Xie S, Xu X, Liang Z, Meng X, Li Q, Zhu Z, Li M, Le XC, Shao Y (2008) Electrochemical DNAzyme sensor for lead based on amplification of DNA–Au bio-bar codes. Ana Chem 80(16):6323–6328. doi:10.1021/ac800601y

71. Hollenstein M, Hipolito C, Lam C, Dietrich D, Perrin DM (2008) A highly selective DNAzyme sensor for mercuric ions. Angew Chem Int Ed 47(23):4346–4350. doi:10.1002/anie.200800960

72. Liu J, Brown AK, Meng X, Cropek DM, Istok JD, Watson DB, Lu Y (2007) A catalytic beacon sensor for uranium with parts-per-trillion sensitivity and millionfold selectivity. Proc Natl Acad Sci U S A 104(7):2056–2061. doi:10.1073/pnas.0607875104

73. Peracchi A (2000) Preferential activation of the 8-17 Deoxyribozyme by Ca^{2+} ions. J Biol Chem 275(16):11693–11697. doi:10.1074/jbc.275.16.11693

74. Georgopoulos PG, Roy A, Yonone-Lioy MJ, Opiekun RE, Lioy PJ (2001) Environmental copper: its dynamics and human exposure issues. J Toxicol Environ Health B Crit Rev 4(4):341–394. doi:10.1080/109374001753146207

75. Chan M-S, Huang S-D (2000) Direct determination of cadmium and copper in seawater using a transversely heated graphite furnace atomic absorption spectrometer with Zeeman-effect background corrector. Talanta 51(2):373–380. doi:10.1016/s0039-9140(99)00283-0

76. Etienne M, Bessiere J, Walcarius A (2001) Voltammetric detection of copper(II) at a carbon paste electrode containing an organically modified silica. Sens Actuators B B76(1–3):531–538. doi:10.1016/S0925-4005(01)00614-1

77. Wu J, Boyle EA (1997) Low blank preconcentration technique for the determination of lead, copper, and cadmium in small-volume seawater samples by isotope dilution ICPMS. Anal Chem 69(13):2464–2470. doi:10.1021/ac961204u

78. Viguier RF, Hulme AN (2006) A sensitized europium complex generated by micromolar concentrations of copper(I): toward the detection of copper(I) in biology. J Am Chem Soc 128(35):11370–11371. doi:10.1021/ja064232v

79. Hong S, Kang T, Moon J, Oh S, Yi J (2007) Surface plasmon resonance analysis of aqueous copper ions with amino-terminated self-assembled monolayers. Colloids Surf A 292(2–3):264–270. doi:10.1016/j.colsurfa.2006.06.031

80. Zheng Y, Gattas-Asfura KM, Konka V, Leblanc RM (2002) A dansylated peptide for the selective detection of copper ions. Chem Commun 20:2350–2351. doi:10.1039/B208012E

81. Gattas-Asfura KM, Leblanc RM (2003) Peptide-coated CdS quantum dots for the optical detection of copper(II) and silver(I). Chem Commun 21:2684–2685. doi:10.1039/B308991F

82. Liu J, Lu Y (2007) A DNAzyme catalytic beacon sensor for paramagnetic Cu^{2+} ions in aqueous solution with high sensitivity and selectivity. J Am Chem Soc 129(32):9838–9839. doi:10.1021/ja0717358

83. Yin BC, Zuo P, Huo H, Zhong X, Ye BC (2010) DNAzyme self-assembled gold nanoparticles for determination of metal ions using fluorescence anisotropy assay. Anal Biochem 401(1):47–52. doi:10.1016/j.ab.2010.02.014

84. Turk B (2006) Targeting proteases: successes, failures and future prospects. Nat Rev Drug Discov 5(9):785–799. doi:10.1038/nrd2092

85. Tong L (2002) Viral proteases. Chem Rev 102(12):4609–4626. doi:10.1021/cr010184f

86. Hedstrom L (2002) Serine protease mechanism and specificity. Chem Rev 102(12): 4501–4524. doi:10.1021/cr000033x

87. Profumo A, Turci M, Damonte G, Ferri F, Magatti D, Cardinali B, Cuniberti C, Rocco M (2003) Kinetics of fibrinopeptide release by thrombin as a function of CaCl₂ concentration: different susceptibility of FPA and FPB and evidence for a fibrinogen isoform-specific effect at physiological Ca^{2+} concentration. Biochemistry 42(42):12335–12348. doi:10.1021/bi034411e

88. Mohanty AK, Simmons CR, Wiener MC (2003) Inhibition of tobacco etch virus protease activity by detergents. Protein Expres Purif 27(1):109–114. doi:10.1016/s1046-5928 (02)00589-2

89. Olsen K, Otte J, Skibsted LH (2000) Steady-state kinetics and thermodynamics of the hydrolysis of β-lactoglobulin by trypsin. J Agric Food Chem 48(8):3086–3089. doi:10.1021/jf991191w

90. Bruchez M, Moronne M, Gin P, Weiss S, Alivisatos AP (1998) Semiconductor nanocrystals as fluorescent biological labels. Science 281(5385):2013–2016. doi:10.1126/science. 281.5385.2013

91. Boeneman K, Mei BC, Dennis AM, Bao G, Deschamps JR, Mattoussi H, Medintz IL (2009) Sensing caspase 3 activity with quantum dot-fluorescent protein assemblies. J Am Chem Soc 131(11):3828–3829. doi:10.1021/ja809721j

92. Guarise C, Pasquato L, De Filippis V, Scrimin P (2006) Gold nanoparticles-based protease assay. Proc Natl Acad Sci U S A 103(11):3978–3982. doi:10.1073/pnas.0509372103

93. Kim Y-P, Oh Y-H, Oh E, Ko S, Han M-K, Kim H-S (2008) Energy transfer-based multiplexed assay of proteases by using gold nanoparticle and quantum dot conjugates on a surface. Anal Chem 80(12):4634–4641. doi:10.1021/ac702416e

94. Medintz IL, Clapp AR, Brunel FM, Tiefenbrunn T, Tetsuo Uyeda H, Chang EL, Deschamps JR, Dawson PE, Mattoussi H (2006) Proteolytic activity monitored by fluorescence resonance energy transfer through quantum-dot-peptide conjugates. Nat Mater 5(7):581–589. doi:10.1038/nmat1676

95. Yin BC, Zhang M, Tan W, Ye BC (2010) Peptide-functionalized spherical polyelectrolyte nanobrushes for real-time sensing of protease activity. ChemBioChem 11(4):494–497. doi:10.1002/cbic.200900735

96. Guo X, Ballauff M (2001) Spherical polyelectrolyte brushes: comparison between annealed and quenched brushes. Phys Rev E 64(5):051406. doi:10.1103/PhysRevE.64.051406

97. Wittemann A, Haupt B, Ballauff M (2003) Adsorption of proteins on spherical polyelectrolyte brushes in aqueous solution. Phys Chem Chem Phys 5(8):1671–1677. doi:10.1039/b300607g

98. Anikin K, Röcker C, Wittemann A, Wiedenmann J, Ballauff M, Nienhaus GU (2005) Polyelectrolyte-mediated protein adsorption: Fluorescent protein binding to individual polyelectrolyte nanospheres. J Phys Chem B 109(12):5418–5420. doi:10.1021/jp0506282

99. Haupt B, Neumann T, Wittemann A, Ballauff M (2005) Activity of enzymes immobilized in colloidal spherical polyelectrolyte brushes. Biomacromolecules 6(2):948–955. doi:10.1021/bm0493584

100. Samokhina L, Schrinner M, Ballauff M, Drechsler M (2007) Binding of oppositely charged surfactants to spherical polyelectrolyte brushes: a study by cryogenic transmission electron microscopy. Langmuir 23(7):3615–3619. doi:10.1021/la063178t

101. Di Nisio M, Middeldorp S, Büller HR (2005) Direct thrombin inhibitors. New Engl J Med 353(10):1028–1040. doi:10.1056/NEJMra044440

102. Ayala Y, Cera ED (1994) Molecular recognition by thrombin. Role of the slows → fast transition, site-specific ion binding energetics and thermodynamic mapping of structural components. J Mol Biol 235(2):733–746. doi:10.1006/jmbi.1994.1024

103. Polsky R, Gill R, Kaganovsky L, Willner I (2006) Nucleic acid-functionalized Pt nanoparticles: catalytic labels for the amplified electrochemical detection of biomolecules. Anal Chem 78(7):2268–2271. doi:10.1021/ac0519864

104. Radi A-E, Acero Sánchez JL, Baldrich E, O'Sullivan CK (2005) Reagentless, reusable, ultrasensitive electrochemical molecular beacon aptasensor. J Am Chem Soc 128(1): 117–124. doi:10.1021/ja053121d

105. Bock LC, Griffin LC, Latham JA, Vermaas EH, Toole JJ (1992) Selection of single-stranded DNA molecules that bind and inhibit human thrombin. Nature 355(6360):564–566. doi:10.1038/355564a0

106. Kim Y, Cao Z, Tan W (2008) Molecular assembly for high-performance bivalent nucleic acid inhibitor. Proc Natl Acad Sci U S A 105(15):5664–5669. doi:10.1073/pnas. 0711803105

107. Tasset DM, Kubik MF, Steiner W (1997) Oligonucleotide inhibitors of human thrombin that bind distinct epitopes. J Mol Biol 272(5):688–698. doi:10.1006/jmbi.1997.1275

Chapter 3
Versatile Graphene-Based Nano-Bio Probe Design and Its Application

3.1 Introduction

Recently there has been an explosion of interest in a new nanomaterial, graphene, which was first discovered in 2004, the isolation from crystalline graphite and characterization of which led to the 2010 Nobel Prize in Physics [1]. Graphene is an allotrope of carbon, the structure of which is one-atom-thick and a two-dimensional (2D) planar sheet of hexagonally arranged sp^2-bonded carbon atoms isolated from its three-dimensional parent material, graphite. Related materials include few-layer-graphene (FLG), ultrathin graphite, graphene oxide (GO), reduced graphene oxide (rGO), and graphene nanosheets (GNS) [2]. This 2D carbon nanomaterial has been attracting much attention due to its excellent electronic, thermal and mechanical features [3–7], as well as its potential applications in synthesizing nanocomposites [8–10], fabricating various micro-electrical devices [11–13], ultrasensitive sensors [14, 15], and drugs delivery [16]. Graphene makes an excellent sensor due to its 2D structure. The fact that its entire volume is exposed to its surrounding makes it very efficient to detect adsorbed molecules [13]. Moreover, graphene is apt to be chemically modified to couple with biomolecules, such as DNA, protein etc. Mohanty et al. utilized this feature as well as graphene's large surface area, atomic-thickness and molecularly-gatable structure to make antibody-functionalized-graphene-sheets, which are excellent candidates for mammalian and microbial detection and diagnosis [17]. Yang's group revealed that GO can act collectively as a quencher for the fluorophore [18].

The marriage of graphene with biomolecules, bioactive compounds, oligomers, and so on provides new opportunities for the sustainable progress of analytical science. Recently, graphene-based analytical systems are a very active area. The literature on the graphene-based analytical systems is growing rapidly, including studies primarily motivated by (1) biochemical applications, and (2) environmental monitoring and safety (EMS). Some recent reviews have summarized the significant developments in synthesis, molecular engineering, thin film, hybrids, energy and

B.-C. Ye et al., *Nano-Bio Probe Design and Its Application for Biochemical Analysis*, Springerbriefs in Molecular Science, DOI: 10.1007/978-3-642-29543-0_3, © The Author(s) 2012

biomedical or analytical applications related to graphene-based nanomarterials, [2, 19–22]. These areas that have been reviewed by other groups will not be covered in detail here. In this chapter, we will present some recent trends in graphene-based bio-analytical systems with an emphasis on nano-bio probe design and its applications.

3.2 Graphene-DNA Molecular Beacon Design and its Application

Molecular beacons (MBs) have been developed by Tyagi and Kramer in 1996 [23]. Typically molecular beacons are dually labeled single-stranded oligonucleotide hybridization probes, which form a stem-and-loop structure to bring a fluorophore and a quencher into close proximity and result in fluorescence quenching of the fluorophore [24, 25]. When the probe encounters a complementary nucleic acid target, it undergoes a conformational change to form a rigid probe/target hybrid, which causes the disruption of the stem-and-loop structure and moves the fluorophore and the quencher away from each other, restoring the fluorescence. MBs possess some combined properties that enable the design of novel and powerful diagnostic assays: (1) they only fluoresce when binding to their targets, (2) they can be labeled with a fluorophore of any desired color, and (3) they are so specific that they can easily discriminate single-nucleotide polymorphisms (SNPs). These unprecedented merits have led to their widespread use in molecular and cellular biology, pathogen detection, and biomedical diagnostics [25–27]. Although MBs have found wide applications, they also have limitations with respect to insufficient sensitivity, false-positive signals, difficult synthesis, and in certain cases difficult selection of fluorophore-quencher pairs [28–30]. To address these problems, researchers have recently combined the self-assembly and specific molecular recognition ability of biomolecules with the particular structural and photophysical features of inorganic nanomaterials, such as quantum dots (QDs), gold nanoparticles (AuNPs), and carbon nanotubes (CNTs) to develop new fluorescent probe tools. The inorganic nanomaterials have been successfully used to construct MB probes for the detection of nucleic acids and proteins [31–33]. Mirkin et al. reported a novel intercellular AuNPs-based probe that combines cellular transfection, enzymatic protection, and RNA detection and quantification [32]. Ozkan and coworkers developed multicolor hybrid DNA probes employing QDs conjugated MBs [34]. Tan et al. employed CNTs as a new class of universal quenchers for molecular sensing [29, 30].

Recently, GO has been attracting much attention due to its excellent electronic, thermal, mechanical and photophysical features. Especially, its photophysical feature has made GO a powerful dye quencher [35], thus enabling its application in bioanalytical sensing where MBs with an "on/off" switching design would be highly desirable. Two key features of GO can be exploited in this type of design. First, GO can suitably serve as a "nanoscaffold" for single-stranded DNA

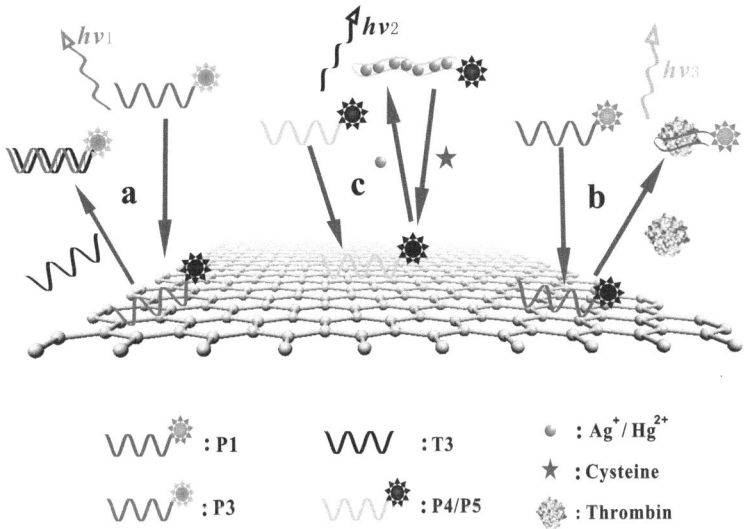

 : P1   \/\/\/ : T3   ● : Ag⁺ / Hg²⁺

 : P3   \/\/\/ : P4/P5   ★ : Cysteine

                                                     ● : Thrombin

Fig. 3.1 Schematic representation of the ssDNA-GO architecture platform for multiplex targets detection. **a** For DNA. **b** For thrombin. **c** For Ag⁺ or Hg²⁺, and cysteine. Reprinted from Ref. [37] with permission from Elsevier

(ssDNA) because ssDNA can be spontaneously absorbed onto the surface of GO by means of π-stacking interactions between nucleotide bases and the GO sheet [36]. Second, GO can act as a "nanoquencher" for the fluorophore. By coupling these two characteristics, a MB-like probe design in which the ssDNA, which is labeled with fluorophore only at one end, can self-organize onto the surface of GO to form a stable ssDNA-GO architecture, while GO is used as the dye quencher. However, in the presence of a target, competitive binding of the target and GO with the ssDNA forces the dye-labeled ssDNA-GO architecture to undergo a conformational alteration in response to interaction with the target, spontaneously liberating the ssDNA from the surface of GO, which results in the restoration of dye fluorescence. Yang's group for the first time employed water-soluble GO (actually graphene can also be used) as a platform for the sensitive and selective detection of DNA and proteins based on a "turn-on" fluorescent strategy [18]. Later, several groups further developed this "turn-on" technique to detect other targets with different DNA designs. It is important to monitor multiple targets in a homogenous sample in various applications. Recently, our group designed a versatile molecular beacon- (MB-) like probe for the multiplex sensing of targets such as sequence-specific DNA, protein, metal ions, and small molecule compounds based on the self-assembled ssDNA-GO architecture, which gives superior sensitivity and rapid response. The ssDNA-GO architecture probe has been successfully applied in the multiplex detection of sequence-specific DNA, thrombin, Ag⁺, Hg²⁺, and cysteine, and the limit of detection was 1, 5, 20, 5.7, and 60 nM, respectively [37]. Fig. 3.1 shows a typical schematic representation of the ssDNA-GO architecture platform for multiplex targets detection.

Yang's group proved that functionalized nanoscale GO can protect oligonu-cleotides from enzymatic cleavage and efficiently deliver oligonucleotides into cells [38]. It is argued that steric hindrance prevents nucleases from effectively attacking the adsorbed phase DNA. Based on the principle, Yang's group also developed an amplified aptamer-based assay through catalytic recycling of the analyte [39]. As an example, an ATP aptamer labeled with a dye was adsorbed on the GO surface, which can protect aptamers from nuclease cleavage and quench the fluorescence of dye-labeled aptamers due to the excellent electronic transfer-ence of GO. When challenged with a target ATP, the aptamer forms a stable, rigid structure and thus is released from the GO surface; in that case, the nuclease can cleave the free aptamer, liberate the fluorophore and ultimately release the target ATP. The released target ATP then binds to another aptamer, and the cycle starts anew, which leads to significant amplification of the signal. By monitoring the increase in fluorescence intensity, the target ATP could be detected with very high sensitivity [39]. Moreover, Wang et al. designed amplified ATP assay platform for sensitive ATP detection in vivo [40]. Also, similar to Yang's group, Tan's group reported a simple and sensitive aptamer/GO based assay for insulin detection with high sensitivity via the catalytic recycling of the analyte [41]. The detection of insulin is especially significant in clinical work, because the level of insulin is the most critical indicator of the endocrinal function of beta cells and serves as a valuable basis for the diagnosis of diabetes mellitus, insulinoma, insulin resistant syndrome, etc. A rapid and reliable laboratory test for the detection of insulin levels in serum can be helpful to diagnose diabetes in its early stages [42, 43]. Very recently, based on the principle of catalytic recycling of the analyte, Cui et al. developed a cyclic enzymatic amplification method for sensitive miRNA detection in complex biological samples that utilized the GO-based probe [44]. MicroRNAs (miRNAs) are 19–25 nt non-coding RNAs that are processed from longer endogenous hairpin transcripts by the enzyme dicer [45, 46]. Several hundred miRNAs are encoded in the human genome and dozens play roles in a diverse variety of cellular processes, both in normal physiology and in disease [47]. Method for sensitive and selective detection of miRNAs are imperative to miRNA discovery, study and clinical diagnosis [48, 49].

In modern biological analysis, various kinds of organic dyes are extensively used. However, with each passing year, more flexibility is being required of these dyes, and the traditional dyes are often unable to meet the expectations [50]. To this end, QDs, semiconductor fluorescent emitters, have quickly filled in the role, being found to be superior to traditional organic dyes on several counts, such as high quantum yield, narrow, symmetric, and stable fluorescence, and size-dependent and tunable absorption and emission [51, 52]. The broad absorption and narrow emission spectra of the QDs have made them excellent donors of fluo-rescence resonance energy transfer (FRET). In this respect, Dong et al. designed a novel strategy for effective sensing of different biomolecules such as DNA and thrombin, etc. based on FRET from QDs to GO through employing an inexpensive MB-functionalized QD instead of dye-labeled MB [53]. This strategy has opened new opportunities for the design of more novel GO-based MB probes. Moreover,

Liu et al. recently reported a label-free fluorescent sensor for the detection of Cu^{2+} based on DNA cleavage-dependent GO-quenched DNAzymes, in which GelRed (a chemical dye) and GO act as the fluorophore and quencher, respectively. This design has further broadened the range of applications for GO-based molecular beacons. Compared with other label-free DNAzyme-based fluorescent platforms, this novel strategy not only maintains the high catalytic activity of Cu^{2+}-dependent DNAzyme, but also effectively improves the fluorescence signal [54].

Understanding and precisely controlling the binding at nano-bio interface is important in basic surface science research, biomedical engineering, biosensor development and nanotechnology. Huang et al. demonstrated the synergistic pH effect on the binding of the aptamer to its target and to GO [55]. They systemically studied that the GO/aptamer system can be reversibly operated by changing the solution pH. At lower pH, the aptamer/GO binding is enhanced while aptamer/target binding is weakened, making this system a regenerable biosensor without covalent conjugation. This method can work for the detection of small molecules and metal ion targets. Moreover, they showed the dual control of the aptamer adsorption allowed the use of this system as a molecular logic gate (an "AND" gate).

3.3 Graphene-Protein Nanoprobe Design and its Application

Biosensors based on the π-stacking interaction between DNA and GO has been investigated to detect various targets (see Sect. 3.2). The marriage of graphene with other biomolecules to develop novel biosensors with versatile features also possesses great potential. To understand biophysicochemical interactions at the nano-bio interface can provide great insights into the design of nano-bio probes. Over the past decades, protein adsorption onto nanomaterial surfaces is believed to mediate cell uptake and thus toxic responses, and has drawn increasing attention (reviewed by Nel et al. [56]). Compared to DNA, there is little information on the interplay of graphene and proteins or graphene-amnio-acid interactions and graphene-protein-based nano-bio probe design. The high specific surface area potentially endows graphene with larger protein adsorption capacities relative to most other nanomaterials.

Recently, our group designed a novel GO-based biosensing platform using peptides as probe biomolecules to specifically recognize the target, and demonstrated its application for the protease activity assay based on FRET between GO (acceptor) and dye-labeled peptide (donor) [57]. To the best of our knowledge, this is the first work of GO-based biosensing platform with a non-DNA probe, and opens new opportunities for designing more novel sensing strategies. Fig. 3.2 shows the principle of the FRET-based peptide-GO biosensor proposed in this work.

It is known that GO is an amphiphile with hydrophilic ionizable edges (-COOH group) and a more hydrophobic basal plane [58]. Once the dye-labeled peptide is

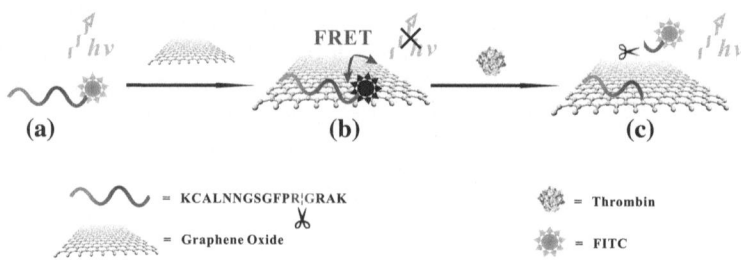

Fig. 3.2 Schematic illustration of the peptide-GO bioconjugate as a sensing probe to monitor the proteolytic activity of thrombin. The peptide was labeled with fluorescein isothiocyanate (*FITC*). The cleavage point at Arg-Gly bonds is indicated by a dashed line. Reprinted from Ref. [57] by permission of the Royal Society of Chemistry

incubated with GO nanosheet, it can undergo stacking interactions with a largely hydrophobic basal phane of GO via aromatic and hydrophobic residues, and also show electrostatic function with ionizable edges via charged and polar residues. As a result, the fluorescence of the resultant peptide-GO self-assembled bioconjugate is quenched immediately and greatly due to the strong energy-transfer interactions between the dye molecule and the GO nanosheet (Fig. 3.2b). In the presence of the studied protease, the peptide hydrolysis on GO nanosheet could occur and subsequently release dye-labeled peptide segments from GO, and then turn on the fluorescence from the default off state due to quenching (Fig. 3.2c). Thereby the protease-catalyzed peptide-GO bioconjugate hydrolysis event is translated into measurable fluorescence enhancement via the free-cleaved dye-labeled peptide segments, which, in turn, results in real-time monitoring of protease activity. This design takes advantage of two key features of the GO. First, GO is suitably served as a "nanoscaffold" for the peptide, because the peptide can be spontaneously absorbed onto the surface of the GO via the special amino acid residues. Second, GO can work as a "nanoquencher" for the dye molecule that is conjugated to the peptide.

To realize such a novel GO-based peptide biosensor design, we should understand the absorption mechanism of the interaction between the peptide and the GO nanosheet. Some attempts were done to illustrate that peptide-GO interaction would work via the driving forces from electrostatic interaction or from specific amino acid residues affinity. We first investigated the interaction between GO with different amino acids. The results showed that only five amino acids exhibited the absorption phenomena: three amino acids of lysine (Lys), histidine (His) and arginine (Arg) with positively charged side chains; and two amino acids of tyrosine (Tyr) and phenylalanine (Phe) with hydrophobic side chains. It was observed that the binding strength between the amino acids and GO follows the order Arg > His > Lys > Tyr > Phe. The –COOH groups in the GO edges (*pKa*: ca. 5) were ionizable and negatively charged in our experimental condition. Thus GO absorbed with the amino acids of Lys, His, and Arg with positively charged side chains via electrostatic interaction as a driving force. Tyr has a special role by virtue of the phenol functionality, which can acts as an electron donor in the reduction of oxides through the deprotonation of its phenolic OH-group. Epoxides in GO edges are reactive. Based on the above facts, the absorption of Tyr onto the GO can be attributed to the interaction between phenolic groups of Tyr and epoxides of GO. This result was in good agreement with the recent study of Deng's group [59]. They developed a general protein-conjugated GO platform to efficiently assemble the nanoparticles using bovine serum albumin (BSA) as reductant via the Tyr residues in BSA. In addition, a theoretical study on the interaction of aromatic amino acids with graphene and SWNTs demonstrated that the aromatic rings of the amino acids (*e.g.* His, Tyr and Phe) prefer to orient in parallel with respect to the plane of graphene via the weak π-π interactions [60]. This helps to explain the reaction between His, Tyr, Phe and GO nanosheet. Moreover, we used His-rich peptide-functionalized gold nanoparticles (AuNPs) for a color indicator to investigate the binding force between His and GO via agarose gel electrophoresis assay. A conclusion can be drawn that His interacts with GO by not only electrostatic but also π-π interaction, which was consistent with the above amino acid analysis.

Based on the above fundamental measurements, a thrombin-recognizing peptide was designed to test the general feasibility of this GO-peptide biosensing platform for the protease activity assay [57]. The development of rapid and simple methods for screening chemical libraries for novel inhibitors of proteases is of great importance in the pharmaceutical industry. The GO-peptide-based FRET assay was also investigated for monitoring the inhibition of thrombin activity in the presence of inhibitors. This novel GO-peptide probe design offers many advantages, including simplicity of preparation and manipulation compared with other methods that employ specific strategies. Moreover, the concept can easily be extended to construct a series of probes by simply changing the peptide substrate sequence to monitor the hydrolysis activity of other proteases and kinases.

Similarly, Ma's group applied this GO-peptide-based biosensing platform to matrix metalloproteinase 2 (MMP2) sensing using a FITC-labeled peptide that contains the core substrate of MMP2, and its applicability has been demonstrated by monitoring the concentration of MMP2 secreted by HeLa cells [61]. Wang et al. developed a sensitive, simple, and robust intracellular protease sensor for high-contrast imaging of apoptotic signaling in live cells based on intracellular delivery of GO-peptide conjugate and caspase-3-mediated cleavage of the substrate peptide [62]. Compared with the above prevailing strategy of noncovalent assembly for graphene-based biosensor construction, the novel strategy of using of a covalent conjugate proposed by Wang et al. can eliminate effectively the displacement of peptide probes caused by protein adsorption, thus substantially improving the stability of protease sensors in protein-abundant media.

Apart from the fluorescence quenching properties of GO toward fluorescent dyes, it is also reported that GO has intrinsic photoluminescence [63]. Jung et al. reported a GO-based immuno-biosensor for detecting a rotavirus as a pathogen model via GO photoluminescence quenching induced by fluorescence resonance energy transfer (FRET) between GO sheets and AuNPs [64]. The antibodies for rotavirus are immobilized on the GO array by a classical EDC/NHS reaction, and the rotavirus cell is captured by specific antigen–antibody interaction. The capture of a target cell was verified by observing the fluorescence quenching of GO by FRET between the GO and AuNPs.

3.4 Other Graphene-Based Nanoprobes and Their Applications

Several GO-based biosensors have been developed by taking advantage of strong adsorption of dye-labeled ssDNA or peptide onto the GO surface and subsequent efficient fluorescence quenching of the dyes. As reviewed in previous sections, in these GO-based sensors, π-π stacking, hydrophobic, and/or electrostatic interactions play important roles in bringing GO and the dye-labeled probe into close proximity to quench the dye fluorescence via resonance energy transfer. Additionally, GO photoluminescence can also be quenched by FRET between GO sheets and AuNPs. So far, the GO-based biosensors have been successfully applied to detect ssDNA, metal ions, helices, small molecules, insulin, miRNAs, proteases, pathogens and living cells. These sensors are strongly dependent on commercially available dye-labeled DNA, peptide or protein as the probe, or quantum dots-modified probes, as a consequence, their detecting targets are constrained due to the limited biorecognition capability of these probes. In this perspective, it is highly desirable to develop more general schemes to expand the target range for GO-based sensing.

Direct π-π stacking and electrostatic interactions between GO and fluorescent dyes or chromophores provide great potential for the development of label-free sensors. Cai et al. reported the design and synthesis of a water-soluble pyrene-based butterfly-shaped conjugated oligoelectrolyte (TFP) and its integration with GO for label-free light-up visual detection of heparin [65]. Efficient fluorescence quenching occurred between TFP and GO because of strong electrostatic and π-π stacking interactions, leading to nearly dark emission in the absence of analytes. Addition of heparin to TFP solution significantly minimized the fluorescence quenching of GO toward TFP, which allowed for light-up visual discrimination of heparin from its analogues. This study demonstrated a new synthetic strategy to develop GO-based chemical and biological sensors without the employment of dye-labeled biomolecules.

Similarly, Wang et al. reported a GO-based bioassay using a conjugated oligomer as the water-soluble neutral probe for "turn on" fluorescent detection of Concanavalin A (ConA) and *Escherichia coli* (*E. coli*) [66]. They developed a conjugated oligomer, which possess a high density of α-mannose side chains and a relatively short backbone to increase water-solubility. By virtue of the fluorescence quenching capability of GO, the background fluorescence was low, which ultimately led to efficient visual detection of ConA and *E. coli* with high sensitivity and selectivity.

References

1. Novoselov KS, Geim AK, Morozov SV, Jiang D, Zhang Y, Dubonos SV, Grigorieva IV, Firsov AA (2004) Electric field effect in atomically thin carbon films. Science 306(5696):666–669. doi:10.1126/science.1102896
2. Sanchez VC, Jachak A, Hurt RH, Kane AB (2011) Biological interactions of graphene-family nanomaterials - an interdisciplinary review. Chem Res Toxicol. doi:10.1021/tx200339h
3. de Heer WA, Berger C, Song ZM, Li TB, Li XB, Ogbazghi AY, Feng R, Dai ZT, Marchenkov AN, Conrad EH, First PN (2004) Ultrathin epitaxial graphite: 2D electron gas properties and a route toward graphene-based nanoelectronics. J Phys Chem B 108(52):19912–19916. doi:10.1021/Jp040650f
4. Geim AK, Novoselov KS (2007) The rise of graphene. Nat Mater 6(3):183–191. doi:10.1038/nmat1849
5. Li D, Kaner RB (2008) Materials science. Graphene-based materials. Science 320(5880):1170–1171. doi:10.1126/science.1158180
6. Westervelt RM (2008) Applied physics. Graphene nanoelectronics. Science 320(5874):324–325. doi:10.1126/science.1156936
7. Freitag M (2008) Graphene: nanoelectronics goes flat out. Nat Nanotechnol 3(8):455–457. doi:10.1038/nnano.2008.219
8. Kamat PV, Muszynski R, Seger B (2008) Decorating graphene sheets with gold nanoparticles. J Phys Chem C 112(14):5263–5266. doi:10.1021/Jp800977b
9. Stankovich S, Dikin DA, Dommett GH, Kohlhaas KM, Zimney EJ, Stach EA, Piner RD, Nguyen ST, Ruoff RS (2006) Graphene-based composite materials. Nature 442(7100):282–286. doi:10.1038/nature04969
10. Williams G, Seger B, Kamat PV (2008) TiO2-graphene nanocomposites. UV-assisted photocatalytic reduction of graphene oxide. ACS Nano 2 (7):1487–1491 doi:10.1021/nn800251f
11. Bunch JS, van der Zande AM, Verbridge SS, Frank IW, Tanenbaum DM, Parpia JM, Craighead HG, McEuen PL (2007) Electromechanical resonators from graphene sheets. Science 315(5811):490–493. doi:10.1126/science.1136836
12. Robinson JT, Perkins FK, Snow ES, Wei Z, Sheehan PE (2008) Reduced graphene oxide molecular sensors. Nano Lett 8(10):3137–3140. doi:10.1021/nl8013007
13. Schedin F, Geim AK, Morozov SV, Hill EW, Blake P, Katsnelson MI, Novoselov KS (2007) Detection of individual gas molecules adsorbed on graphene. Nat Mater 6(9):652–655. doi:10.1038/nmat1967
14. Zhou M, Zhai Y, Dong S (2009) Electrochemical sensing and biosensing platform based on chemically reduced graphene oxide. Anal Chem 81(14):5603–5613. doi:10.1021/ac900136z
15. Zuo X, He S, Li D, Peng C, Huang Q, Song S, Fan C (2010) Graphene oxide-facilitated electron transfer of metalloproteins at electrode surfaces. Langmuir 26(3):1936–1939. doi:10.1021/la902496u
16. Liu Z, Robinson JT, Sun X, Dai H (2008) PEGylated nanographene oxide for delivery of water-insoluble cancer drugs. J Am Chem Soc 130(33):10876–10877. doi:10.1021/ja803688x
17. Mohanty N, Berry V (2008) Graphene-based single-bacterium resolution biodevice and DNA transistor: interfacing graphene derivatives with nanoscale and microscale biocomponents. Nano Lett 8(12):4469–4476. doi:10.1021/nl802412n
18. Lu CH, Yang HH, Zhu CL, Chen X, Chen GN (2009) A graphene platform for sensing biomolecules. Angew Chem Int Ed Engl 48(26):4785–4787. doi:10.1002/anie.200901479
19. Feng L, Liu Z (2011) Graphene in biomedicine: opportunities and challenges. Nanomedicine (Lond) 6(2):317–324. doi:10.2217/nnm.10.158
20. Guo S, Dong S (2011) Graphene nanosheet: synthesis, molecular engineering, thin film, hybrids, and energy and analytical applications. Chem Soc Rev 40(5):2644–2672. doi:10.1039/c0cs00079e

21. Kuila T, Bose S, Khanra P, Mishra AK, Kim NH, Lee JH (2011) Recent advances in graphene-based biosensors. Biosens Bioelectron 26(12):4637–4648. doi:10.1016/j.bios.2011.05.039
22. Wang Y, Li Z, Wang J, Li J, Lin Y (2011) Graphene and graphene oxide: biofunctionalization and applications in biotechnology. Trends Biotechnol 29(5):205–212. doi:10.1016/j.tibtech.2011.01.008
23. Tyagi S, Kramer FR (1996) Molecular beacons: probes that fluoresce upon hybridization. Nat Biotechnol 14(3):303–308. doi:10.1038/nbt0396-303
24. Tan W, Wang K, Drake TJ (2004) Molecular beacons. Curr Opin Chem Biol 8(5):547–553. doi:10.1016/j.cbpa.2004.08.010
25. Venkatesan N, Seo YJ, Kim BH (2008) Quencher-free molecular beacons: a new strategy in fluorescence based nucleic acid analysis. Chem Soc Rev 37(4):648–663. doi:10.1039/b705468h
26. Wang K, Tang Z, Yang CJ, Kim Y, Fang X, Li W, Wu Y, Medley CD, Cao Z, Li J, Colon P, Lin H, Tan W (2009) Molecular engineering of DNA: molecular beacons. Angew Chem Int Ed Engl 48(5):856–870. doi:10.1002/anie.200800370
27. Tyagi S, Bratu DP, Kramer FR (1998) Multicolor molecular beacons for allele discrimination. Nat Biotechnol 16(1):49–53. doi:10.1038/nbt0198-49
28. Song S, Liang Z, Zhang J, Wang L, Li G, Fan C (2009) Gold-nanoparticle-based multicolor nanobeacons for sequence-specific DNA analysis. Angew Chem Int Ed Engl 48(46):8670–8674. doi:10.1002/anie.200901887
29. Yang R, Jin J, Chen Y, Shao N, Kang H, Xiao Z, Tang Z, Wu Y, Zhu Z, Tan W (2008) Carbon nanotube-quenched fluorescent oligonucleotides: probes that fluoresce upon hybridization. J Am Chem Soc 130(26):8351–8358. doi:10.1021/ja800604z
30. Yang R, Tang Z, Yan J, Kang H, Kim Y, Zhu Z, Tan W (2008) Noncovalent assembly of carbon nanotubes and single-stranded DNA: an effective sensing platform for probing biomolecular interactions. Anal Chem 80(19):7408–7413. doi:10.1021/ac801118p
31. Dubertret B, Calame M, Libchaber AJ (2001) Single-mismatch detection using gold-quenched fluorescent oligonucleotides. Nat Biotechnol 19(4):365–370. doi:10.1038/86762
32. Seferos DS, Giljohann DA, Hill HD, Prigodich AE, Mirkin CA (2007) Nano-flares: probes for transfection and mRNA detection in living cells. J Am Chem Soc 129(50):15477–15479. doi:10.1021/ja0776529
33. Xue X, Wang F, Liu X (2008) One-step, room temperature, colorimetric detection of mercury (Hg^{2+}) using DNA/nanoparticle conjugates. J Am Chem Soc 130(11):3244–3245. doi:10.1021/ja076716c
34. Ozkan M, Kim JH, Chaudhary S (2007) Multicolour hybrid nanoprobes of molecular beacon conjugated quantum dots: FRET and gel electrophoresis assisted target DNA detection. Nanotechnology 18(19):195105. doi:10.1088/0957-4484/18/19/195105
35. Swathi RS, Sebastian KL (2009) Long range resonance energy transfer from a dye molecule to graphene has (distance)(-4) dependence. J Chem Phys 130(8):086101. doi:10.1063/1.3077292
36. Chang H, Tang L, Wang Y, Jiang J, Li J (2010) Graphene fluorescence resonance energy transfer aptasensor for the thrombin detection. Anal Chem 82(6):2341–2346. doi:10.1021/ac9025384
37. Zhang M, Yin BC, Tan W, Ye BC (2011) A versatile graphene-based fluorescence "on/off" switch for multiplex detection of various targets. Biosens Bioelectron 26(7):3260–3265. doi:10.1016/j.bios.2010.12.037
38. Lu CH, Zhu CL, Li J, Liu JJ, Chen X, Yang HH (2010) Using graphene to protect DNA from cleavage during cellular delivery. Chem Commun (Camb) 46(18):3116–3118. doi:10.1039/b926893f
39. Lu CH, Li J, Lin MH, Wang YW, Yang HH, Chen X, Chen GN (2010) Amplified aptamer-based assay through catalytic recycling of the analyte. Angew Chem Int Ed Engl 49(45):8454–8457. doi:10.1002/anie.201002822

40. Wang Y, Li Z, Hu D, Lin CT, Li J, Lin Y (2010) Aptamer/graphene oxide nanocomplex for in situ molecular probing in living cells. J Am Chem Soc 132(27):9274–9276. doi:10.1021/ja103169v

41. Pu Y, Zhu Z, Han D, Liu H, Liu J, Liao J, Zhang K, Tan W (2011) Insulin-binding aptamer-conjugated graphene oxide for insulin detection. Analyst 136(20):4138–4140. doi:10.1039/c1an15407a

42. Preethi BL, Jaisri G, Kumar KM, Sharma R (2011) Assessment of insulin resistance in normoglycemic young adults. Fiziol Cheloveka 37(1):118–125. doi:10.1134/S0362119711010154

43. van Belle TL, Coppieters KT, von Herrath MG (2011) Type 1 diabetes: etiology, immunology, and therapeutic strategies. Physiol Rev 91(1):79–118. doi:10.1152/physrev.00003.2010

44. Cui L, Lin X, Lin N, Song Y, Zhu Z, Chen X, Yang CJ (2011) Graphene oxide-protected DNA probes for multiplex microRNA analysis in complex biological samples based on a cyclic enzymatic amplification method. Chem Commun (Camb). doi:10.1039/c1cc15412e

45. Bartel DP (2004) MicroRNAs: genomics, biogenesis, mechanism, and function. Cell 116(2):281–297. doi:S0092867404000455

46. Farh KK, Grimson A, Jan C, Lewis BP, Johnston WK, Lim LP, Burge CB, Bartel DP (2005) The widespread impact of mammalian MicroRNAs on mRNA repression and evolution. Science 310(5755):1817–1821. doi:10.1126/science.1121158

47. Buchan JR, Parker R (2007) Molecular biology. The two faces of miRNA. Science 318(5858):1877–1878. doi:10.1126/science.1152623

48. Arenz C (2006) MicroRNAs–future drug targets? Angew Chem Int Ed Engl 45(31):5048–5050. doi:10.1002/anie.200601537

49. Wark AW, Lee HJ, Corn RM (2008) Multiplexed detection methods for profiling microRNA expression in biological samples. Angew Chem Int Ed Engl 47(4):644–652. doi:10.1002/anie.200702450

50. Walling MA, Novak JA, Shepard JR (2009) Quantum dots for live cell and in vivo imaging. Int J Mol Sci 10(2):441–491. doi:10.3390/ijms10020441

51. Michalet X, Pinaud FF, Bentolila LA, Tsay JM, Doose S, Li JJ, Sundaresan G, Wu AM, Gambhir SS, Weiss S (2005) Quantum dots for live cells, in vivo imaging, and diagnostics. Science 307(5709):538–544. doi:10.1126/science.1104274

52. Wu X, Liu H, Liu J, Haley KN, Treadway JA, Larson JP, Ge N, Peale F, Bruchez MP (2003) Immunofluorescent labeling of cancer marker Her2 and other cellular targets with semiconductor quantum dots. Nat Biotechnol 21(1):41–46. doi:10.1038/nbt764

53. Dong H, Gao W, Yan F, Ji H, Ju H (2010) Fluorescence resonance energy transfer between quantum dots and graphene oxide for sensing biomolecules. Anal Chem 82(13):5511–5517. doi:10.1021/ac100852z

54. Liu M, Zhao H, Chen S, Yu H, Zhang Y, Quan X (2011) Label-free fluorescent detection of Cu(II) ions based on DNA cleavage-dependent graphene-quenched DNAzymes. Chem Commun (Camb) 47(27):7749–7751. doi:10.1039/c1cc12006a

55. Liu JW, Huang PJJ, Kempaiah R (2011) Synergistic pH effect for reversible shuttling aptamer-based biosensors between graphene oxide and target molecules. J Mater Chem 21(25):8991–8993. doi:10.1039/C1jm11702e

56. Nel AE, Madler L, Velegol D, Xia T, Hoek EM, Somasundaran P, Klaessig F, Castranova V, Thompson M (2009) Understanding biophysicochemical interactions at the nano-bio interface. Nat Mater 8(7):543–557. doi:10.1038/nmat2442

57. Zhang M, Yin BC, Wang XF, Ye BC (2011) Interaction of peptides with graphene oxide and its application for real-time monitoring of protease activity. Chem Commun (Camb) 47(8):2399–2401. doi:10.1039/c0cc04887a

58. Kim J, Cote LJ, Kim F, Yuan W, Shull KR, Huang J (2010) Graphene oxide sheets at interfaces. J Am Chem Soc 132(23):8180–8186. doi:10.1021/ja102777p

59. Liu J, Fu S, Yuan B, Li Y, Deng Z (2010) Toward a universal "adhesive nanosheet" for the assembly of multiple nanoparticles based on a protein-induced reduction/decoration of graphene oxide. J Am Chem Soc 132(21):7279–7281. doi:10.1021/ja100938r

60. Rajesh C, Majumder C, Mizuseki H, Kawazoe Y (2009) A theoretical study on the interaction of aromatic amino acids with graphene and single walled carbon nanotube. J Chem Phys 130(12):124911. doi:10.1063/1.3079096

61. Feng D, Zhang Y, Feng T, Shi W, Li X, Ma H (2011) A graphene oxide-peptide fluorescence sensor tailor-made for simple and sensitive detection of matrix metalloproteinase 2. Chem Commun (Camb) 47(38):10680–10682. doi:10.1039/c1cc13975d

62. Wang H, Zhang Q, Chu X, Chen T, Ge J, Yu R (2011) Graphene oxide-peptide conjugate as an intracellular protease sensor for caspase-3 activation imaging in live cells. Angew Chem Int Ed Engl 50(31):7065–7069. doi:10.1002/anie.201101351

63. Sun X, Liu Z, Welsher K, Robinson JT, Goodwin A, Zaric S, Dai H (2008) Nano-graphene oxide for cellular imaging and drug delivery. Nano Res 1(3):203–212. doi:10.1007/s12274-008-8021-8

64. Jung JH, Cheon DS, Liu F, Lee KB, Seo TS (2010) A graphene oxide based immuno-biosensor for pathogen detection. Angew Chem Int Ed Engl 49(33):5708–5711. doi:10.1002/anie.201001428

65. Cai L, Zhan R, Pu KY, Qi X, Zhang H, Huang W, Liu B (2011) Butterfly-shaped conjugated oligoelectrolyte/graphene oxide integrated assay for light-up visual detection of heparin. Anal Chem. doi:10.1021/ac2016135

66. Wang L, Pu KY, Li J, Qi X, Li H, Zhang H, Fan C, Liu B (2011) A graphene-conjugated oligomer hybrid probe for light-up sensing of lectin and Escherichia coli. Adv Mater 23(38):4386–4391. doi:10.1002/adma.201102227

Chapter 4
Metal Nanoparticles-Based Colorimetric Probe Design and Its Application

4.1 Introduction

Molecular recognition plays a vital role in biological systems and can be often observed in between receptor-ligand, antigen–antibody, DNA–protein, RNA-ribosome, etc. Assays that enable rapid, sensitive, high-throughput monitoring of molecule recognition are also demonstrated to be important in molecular/cellular biology, drug development, biomedical diagnostics, food safety, environmental monitoring, forensic analysis, and civil defence [1–3]. Biosensors, combining the high selectivity of biomolecule recognition and the high sensitivity of advanced optoelectronic transducers, are developed to effectively convert biological recognition events into measurable physical or chemical signals. The recent advances of nanoscience and nanotechnology open new opportunities for the application of nanomaterials as signal transducers in biosensor-based assays. Metal nanoparticles (metal NPs), such as AuNPs and AgNPs, are particularly useful in this regard due to many unique properties. The metal NPs-based colorimetric sensors have been widely developed for detecting various analytes including oligonucleotides [4, 5], small molecules [6–10], proteins [11, 12], carbohydrates [13, 14], metal ions [15–17], inorganic anions [18–20] etc. in homogeneous solutions, which depend on the targeted analyte-induced reversible color switch between dispersion-aggregation states of chemically functionalized NPs. The metal NPs-based colorimetric method can be easily monitored by the naked eye without advanced instruments, which is increasingly becoming a routine bioassay that could provide comparable or even better sensitivity and selectivity than conventional molecular fluorescent assays. Metal NPs with sizes of 1–100 nm possess unique features like high surface-to-volume ratio [21–23], quantum size effect [24, 25], and electrodynamic interactions [26, 27], which are responsible for observed optical, electronic, chemical, and magnetic properties. In this chapter, we highlight the marriage of biomolecule recognition with metal NPs (AuNPs and AgNPs) for colorimetric probe design and its applications.

B.-C. Ye et al., *Nano-Bio Probe Design and Its Application for Biochemical Analysis*,
Springer Briefs in Molecular Science, DOI: 10.1007/978-3-642-29543-0_4,
© The Author(s) 2012

4.2 Sensitive and Rapid Colorimetric Detection of Biological Thiols Using Nucleotide-Stabilized Silver Nanoparticles

Biological thiols, such as cysteine (Cys), homocysteine (Hcys) and glutathione (GSH), play vital roles in a variety of important biochemical pathways [28, 29]. They are recognized as important biomarkers for various medical conditions in view of their levels in physiological fluids, such as plasma and urine [30, 31]. Thus, it is highly desirable to develop a simple, rapid, sensitive, and economical routine assay to detect these thiols. Recently, remarkable progresses have been made in the development of colorimetric sensors for detecting cysteine. The Mirkin group employed DNA-AuNPs bioconjugate as a probe for colorimetric detection of cysteine [32]. But their method needs some complicated steps, such as modifying DNA onto the AuNPs. To address this problem, the Li's group utilized AuNPs as a probe for detecting cysteine based on the aggregation of AuNPs induced by the cysteine/Cu^{2+} complex [33]. However, this assay requires Cu^{2+} as a cross-linking agent for AuNPs aggregation.

In view of colorimetric sensing, AgNPs are also of significant importance to biological detection as an alternative of AuNPs. Similarly, AgNPs display a distance-relevant color, and higher extinction coefficient than that of AuNPs in the same size [34], which entitle them to be utilized as ideal color-reporting elements for colorimetric sensor design. AgNPs are usually prepared by reducing $AgNO_3$ with citrate sodium, and the resultant nanoparticles are capped with citrate group. It is well known that nucleic acid can interact with metal ions through nucleobases and the phosphodiester backbone. Adenosine 5′-triphosphate (ATP) consists of adenosine and three phosphate groups, which suggest that it can serve as an effective capping ligand during the nanoparticles growth process: adenine can bind to nanoparticle surfaces due to the interaction between functional groups (amines, carbonyls) of the nucleobases and the metal surface [35, 36], and the negatively charged phosphate group can stabilize the nanoparticles against aggregation in the phase of their growth through electrostatic repulsion. Enlightened by the above facts, we developed a facile and more environmentally friendly method to synthesize well-dispersed AgNPs by using ATP as a capping ligand, and demonstrated their application for biological thiols detection as a color-sensing element.

In this section, we propose a novel strategy for colorimetric detection of bio-thiols using ATP-stabilized AgNPs. Cysteine, a thiol-containing amino acid, can be coupled to the surface of AgNPs by the form of Ag–S bond. At the same time, cysteine may recognize and bind ATP via a proton bridge with the amino group in position 6 on the adenosine ring of ATP [37]. The cysteine nucleophile forms a bond with the phosphorus atom of the phosphate moiety, and the carbonyl of cysteine forms a hydrogen bond with the hydroxyl in the nucleobase of ATP, in that case, a crosslink network forms and induces the aggregation of AgNPs, resulting in appreciable changes in color and absorption properties.

In this work, we synthesized the ATP-capped AgNPs, and tested the performance of the two types of AgNPs (AgNP-1: citrate stabilized; AgNP-2: ATP

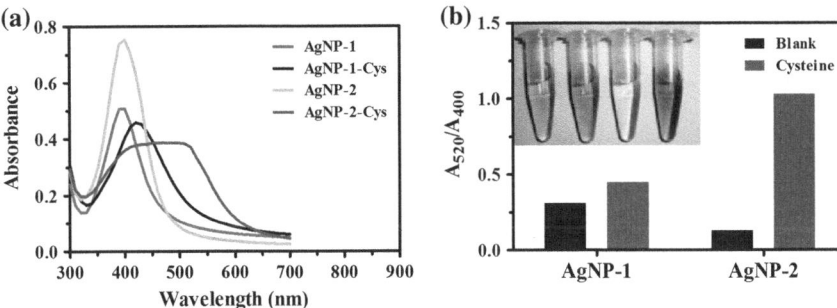

Fig. 4.1 **a** Typical absorption spectra of AgNPs (*AgNP-1* citrate stabilized; *AgNP-2* ATP stabilized) in the absence and presence of 100 μM cysteine; **b** Absorption ratio (A_{520}/A_{400}) and photographs (inset) of AgNPs in the absence and presence of 100 μM cysteine

stabilized) toward the addition of cysteine. Citrate-stabilized AgNPs (AgNP-1) were prepared according to the classical citrate and NaBH$_4$ reduction method. Figure 4.1 illustrates the difference of absorption spectra between the AgNP-1 and AgNP-2 and the color changes upon the addition of cysteine (Fig. 4.1b inset). An absorption peak was observed at 400 nm that originated from the surface plasmon absorption of AgNPs, and upon the addition of cysteine, the absorption peak of AgNP-1/Cys red shifted to 420 nm, while the absorption peak of AgNP-2/Cys red shifted to 520 nm with the color of the solution changed from yellow to red. Cys-induced aggregation of ATP-stabilized AgNPs resulted in a decrease in the plasmon absorption at 400 nm and the formation of a broadened surface plasmon band around 420–520 nm. Considering that the ratios of the absorbance at 520 and 400 nm (A_{520}/A_{400}) are related to the quantities of dispersed and aggregated AgNPs, we chose A_{520}/A_{400} ratio as the indicator of the performance of AgNPs. UV–vis spectroscopy of ATP-stabilized AgNPs showed a smaller A_{520}/A_{400} ratio compared to the citrate-stabilized AgNPs, which indicated the growth of AgNPs using ATP as capping ligand was completed that coupled with a higher A_{400} and a lower A_{520}. Moreover, the spectrum of the ATP-stabilized AgNPs had the narrow full width at half-maximum (FWHM), indicating that the synthesized ATP-stabilized AgNPs were monodispersed and uniform.

Considering the appreciable changes in color and absorption properties of ATP-stabilized AgNPs toward Cys, the potential of developing a novel colorimetric probe for the determination of biothiols was assessed. To quantitatively detect cysteine using ATP-stabilized AgNPs, UV–vis spectra of AgNPs in the absence and presence of different concentrations of Cys were recorded (Fig. 4.2a). It was observed that there were gradual decreases in the short-wavelength band ($\sim$400 nm) whereas there were gradual increases in the long-wavelength band ($\sim$520 nm) with increasing cysteine concentration. The spectrophotometric changes were accompanied by yellow-to-red color changes (Fig. 4.2a inset). From Fig. 4.2b, it can be seen that the absorption ratio (A_{520}/A_{400}) was sensitive to the concentration of cysteine. In Fig. 4.2b inset, the linear range is from 0.01–20 μM

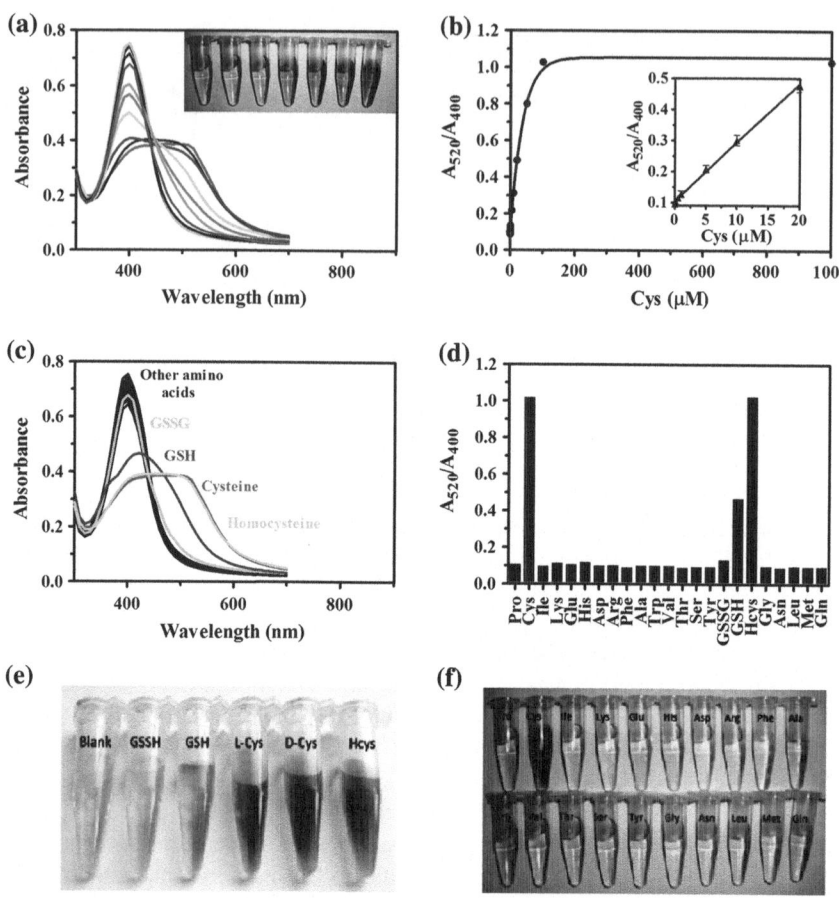

Fig. 4.2 a UV–vis absorbance of ATP-stabilized AgNPs in solution upon addition of cysteine (0, 0.0001, 0.001, 0.01, 0.1, 1, 5, 10, 20, 50, 100, 1000 μM). Inset: photo of ATP-stabilized AgNPs was incubated with varied concentration of cysteine (From *left* to *right*: 0, 0.01, 0.1, 1, 10, 100, 1000 μM); **b** Plot of absorption ratios (A_{520}/A_{400}) corresponding to the varying concentration of cysteine. **c** Typical absorption spectra and **d** Absorption ratio (A_{520}/A_{400}) of ATP-stabilized AgNPs in the absence and presence of 100 μM Cys, Hcys, GSH and various amino acids; **e** and **f** The photographs of ATP-stabilized AgNPs in the presence of 100 μM analytes mentioned above

with linear equation $Y = 0.0195 \, X + 0.1086$, where Y is the absorption ratio (A_{520}/A_{400}) and X is the concentration of Cys (regression coefficient $R^2 = 0.9960$). A distinguishable color change can be monitored upon the ATP-stabilized AgNPs challenged with 100 nM Cys. And the limit of detection (LOD) of Cys based on three times of signal-to-noise level of the blank sample was estimated to be 10.4 nM, which had the similar performance as turn-on fluorescence probes of 10 [38] and 11.4 nM [33]. These data clearly illustrate simple, fast, sensitive, and good linearity for the quantitative analysis of Cys by using the ATP-stabilized

Table 4.1 Determination of the thiols in the human urine samples

Sample	Determined aminothiols (10^{-4} M)	Added cys (10^{-4} M)	Measured (10^{-4} M)	Recovery (%)	RSD (n = 8, %)
1	0.817	1	1.801	98.4	4.0
		5	5.642	96.5	3.9
2	0.681	1	1.644	96.3	6.9
		5	5.551	97.4	5.8
3	0.932	1	1.89	95.8	4.3
		5	5.787	97.1	3.2

AgNPs. Notably, the limit of detection is much lower than the thiols content in plasma (micromolar), which suggests that the present work has great potential for clinic diagnosis.

To test selectivity, competing stimuli including various amino acids each at a concentration of 100 μM was examined under the same conditions as in the case of Cys (Fig. 4.2c, d, e, f). It was found that biological thiols (Cys, Hcys, GSH) result in an obvious change in the A_{520}/A_{400} ratio, while there was nearly negligible A_{520}/A_{400} ratio change when in the presence of other amino acids without thiol groups at the same concentration. Observably, GSH had relatively weaker effect on the aggregation of ATP-stabilized AgNPs compared to Cys and Hcys (about of 50% in the same condition), which might be due to its steric hindrance [39]. However, GSSG, which contains a disulfide bond, had little effect on the aggregation of ATP-stabilized AgNPs.

In order to test the feasibility of our proposed method in real samples, we studied the possible applicability of ATP-stabilized AgNPs for the direct measurment of biothiols in human urine. Generally, aminothiols are present in urine in two forms: reduced thiols characterized by the presence of an –SH function group, impressionable to oxidation and nucleophilic displacement reactions, and oxidized with disulfide bridges ($-S-S-$). The oxidized form encompasses symmetrical and mixed disulfides [40]. After reduction with suitable regents [40], these thiols become free in the urine, which then can be used for analysis. The unknown concentrations of aminothiols in three different urine samples were measured by the standard addition method using cysteine as the standard. The results are listed in Table 4.1, which are in good agreement with those obtained in previous studies [40]. Recovery of added known amount of Cys to the urine samples was in general larger than 95%, which indicated that the present method has a promise in practical application with great accuracy and reliability.

Since all of the nucleosides have similar functional groups (amines, carbonyls, etc.) that could act as ligands for the metallic nanoparticle surface as well as phosphate groups along the backbone that could stabilize nanoparticles via electrostatic repulsion, we further exploited the effects of the ATP analogs (ADP, AMP, GTP, CTP, UTP, dATP, dGTP, dCTP, dTTP, PPPi, and PPi) on the stabilizing AgNPs. There were significantly different A_{520}/A_{400} ratios of the AgNPs formed with the above ATP analogs, following the order AMP > PPi > PPPi > dGTP > dCTP > ADP > CTP > dTTP > UTP > dATP > GTP > ATP. As mentioned above, the A_{520}/A_{400} ratio can be used as a performance of the AgNPs. The lower A_{520}/A_{400} ratio is, the better performance of morpha shows (monodispersed and uniform). Thus, the ATP-stabilized AgNPs displayed the best quality of homogeneity and stability in view of A_{520}/A_{400} ratio. We further discussed that the differences between these prepared AgNPs are mostly attributed to the following factors: the different functional groups (amines and carbonyls) and the different negatively charged phosphate groups present in the nucleotides, namely the more the functional groups and negatively charged phosphate groups that exist, the greater the stability of the formed AgNPs (e.g., ATP > CTP; ATP > ADP > AMP). In addition, as most of the nucleotides chemisorbed on the metal atom through multiple binding sites, the different types of possible surface binding moieties (e.g., carbonyls and amides; mono-versus polydentate) may also affect the observed affinities, as mentioned in the literature [41]. The behaviour of the resultant AgNPs toward cysteine was also tested, upon the cysteine added and incubated for 10 min at RT, the A_{520}/A_{400} of all AgNPs were changed to varying degrees. The ratio of A_{520}/A_{400} in the presence and absence of cysteine was calculated to illustrate the discrimination ability of the AgNPs toward cysteine. The ATP-stabilized AgNPs had the best performance in the discrimination of cysteine in a short time (10 min). And the GTP-stabilized AgNPs had the same performance as ATP-stabilized AgNPs toward cysteine, which is probably due to their structural similarity (consisting of a pyrimidine ring). Moreover, the AgNPs capped with ADP or AMP, which consist of the pyrimidine ring, show the weaker performance than that of ATP, following the sequence ATP > ADP > AMP. The results suggest that the different negatively charged phosphate groups in the nucleotides influence on the stability of AgNPs and the behaviour of cysteine-stimulated AgNPs. We found that the dATP-stabilized AgNPs show less 50% of discrimination ability toward cysteine than that of ATP-stabilized AgNPs. This gave a direct evidence for Cys-stimulated aggregation of the AgNPs that the carbonyl of cysteine forms a hydrogen bond with the hydroxyl in the nucleobase of ATP. Similar facts can also be illustrated from those of GTP and dGTP.

In conclusion, we have developed a novel method to synthesize well-dispersed AgNPs by using adenosine 5′-triphosphate (ATP) as a capping ligand, and utilized this ATP-stabilized AgNPs for highly sensitive and selective colorimetric detection of biological thiols by exploiting the interplasmon coupling in AgNPs. The detection is based on the biothiols-induced aggregation of ATP-stabilized AgNPs. The method allows selective determination of thiols as low as 10.4 nM, which is comparable to the most sensitive method reported for thiols detection. This sensing system was successfully applied to the determination of the total amniothiols in the human urine, which suggested the present work has great potential for clinic diagnosis. We believe that the nucleoside-stabilized AgNPs could be applied as a more general colorimetric probe for sensing other biological analytes of interest.

4.3 Colorimetric Chiral Recognition of Enantiomers Using Nucleotide-Capped Silver Nanoparticles

Chiral recognition is among the important and special modes of molecular recognition, which draws extensive attentions in the field of biochemistry, pharmaceutics, and drug development. The performance of the enantiomers of a chiral molecule may exhibit remarkable discrepancy in terms of biochemical activity, potency, toxicity, transport mechanism, and pathways of metabolism. For instance, L-cysteine plays a vital role in the living system and its deficiency is coupled with a number of clinical symptoms (skin lesions, liver damage, AIDS, and certain neurological disorders), while the role of L-cysteine in the central nervous system is not fully understood [42]. On the other hand, Soutourina et al. reported that D-cysteine could interfere with many targets inside the cell, but the corresponding sites of action of D-cysteine are unknown, and little information is available on the occurrence and roles of D-amino acids [43]. To develop the means of discriminating between enantiomers of a chiral molecule is of critical importance in many fields of analytical chemistry and biotechnology, especially in pharmaceutical science and technology. Therefore, it is highly desirable to develop simple, rapid, sensitive, and high-throughput routine assay for chiral recognition.

It is known that Metal NPs surfaces can exhibit intrinsically chiral structure. Furthermore, chirality can be bestowed onto achiral metal surfaces by adsorption of chiral molecules [44]. Recently, the chirality of these metal NPs has attracted attention, and application in chiral technologies is an interesting perspective. Considerable effort has been devoted to the synthesis and characterization of chiral, optically active ligand-capped metal Nps [45–47]. However, the field of enantioselective recognition using metal NPs still remains unexplored. In this section, we present a simple and reliable colorimetric method for the separation and quantitative determination of enantiomers in aqueous solution using uridine 5′-triphosphate (UTP)-capped AgNPs without any prior derivatization and sample preparation [48].

In this study, as a model system, L- and D-cysteine (Cys) were used to evaluate the UTP-capped AgNPs on the colorimetric discrimination of chiral enantiomers because of their biological importance and their inherent chiral structure [31, 42, 49, 50]. In the presence of D-Cys, an appreciable yellow-to-red color shift of UTP-capped AgNPs can be observed. However, no color changes were found in the presence of L-Cys. More importantly, UTP-capped AgNPs selectively interacted with one enantiomer of cysteine from a solution of racemic cysteine, leaving an excess of the other enantiomer in the solution after centrifugation treatment, thus resulting in enantioselective separation.

The enantioselective interaction of chiral cysteine with UTP-capped AgNPs can also be probed using UV–vis absorbance spectroscopy. An absorption peak was observed at 400 nm that originated from the surface plasmon absorption of Ag-NPs, upon the addition of D-cysteine, the absorption peak red-shifted to 520 nm (Fig. 4.3c), and the color of the solution changed from yellow to red (Fig. 4.4a).

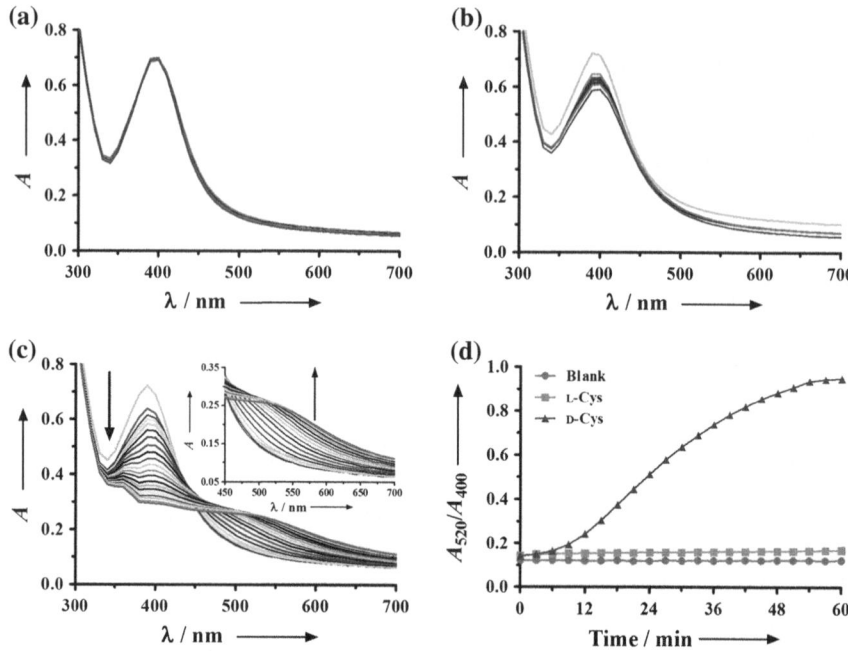

Fig. 4.3 Time-course study on UV–vis absorbance of UTP-capped AgNPs in solution upon addition of **a** H_2O as a blank, **b** 100 μM L-Cys, and **c** 100 μM D-Cys; **d** Plots of absorption ratios (A_{520}/A_{400}) corresponding to H_2O, 100 μM L-Cys, and 100 μM D-Cys. Reprinted with permission from ref [48]. Copyright 2011 American Chemical Society

D-Cys-induced aggregation of UTP-capped AgNPs resulted in a decrease in the plasmon absorption at 400 nm and the formation of a broadened surface plasmon band around 450–600 nm. No L-Cys-induced aggregation was found, as evidenced by the fact that there is no color change (Fig. 4.4a) and significant UV–vis spectral shift (Fig. 4.3b). Fig. 4.3d illustrates the significant difference in the absorbance ratio (A_{520}/A_{400}) of UTP-capped AgNPs responding to L-and D-cysteine over periods of 60 min.

The solutions of UTP-capped AgNPs mixed with L- or D- Cys (100 μM, 60 min) were centrifuged at 12,000 rpm for 5 min, a lot of precipitate can be collected at the bottom of tube which contains UTP-capped AgNPs and D-Cys, however, no obvious changes occurred in the L-Cys tube under the same conditions (Fig. 4.4a, b). To test the feasibility of the UTP-capped AgNPs as enantiospecific adsorbents, the resultant supernatant in D-Cys tube was again interacted with UTP-capped AgNPs. In this case, no colorimetric change was observed, and the significant UV–vis spectral shift (A_{520}/A_{400}) was also not detectable (Fig. 4.4c). The results in Fig. 4.4 suggest that D-cysteine can selectively induce the aggregation of UTP-capped AgNPs, and can be precipitated with AgNPs via the interfacial encapsulation and crosslinking reaction. The work revealed the potential of UTP-capped AgNPs to serve as enantiospecific adsorbents of chiral

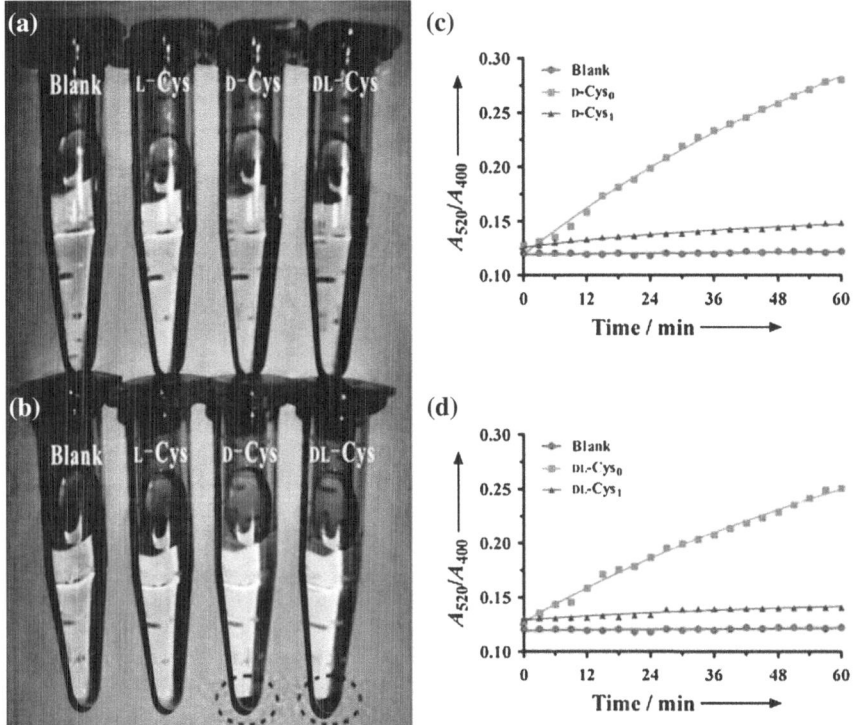

Fig. 4.4   A photo exhibition of UTP-capped AgNPs toward 100 μM L-Cys, D-Cys, and DL-Cys **a** before centrifugation, **b** after centrifugation. Plots of absorption ratios (A_{520}/A_{400}) corresponding to **c** 10 μM D-Cys$_0$, and D-Cys$_1$ (the supernatant of 100 μM D-Cys$_0$ reacted with UTP-capped AgNPs), **d** 10 μM DL-Cys$_0$, and DL-Cys$_1$(the supernatant of 100 μM DL-Cys$_0$ reacted with UTP-capped AgNPs). Reprinted with permission from ref [48]. Copyright 2011 American Chemical Society

species. To further confirm an enantioselective separation and enantiomeric purification of the cysteine utilizing AgNPs, we applied UTP-capped AgNPs to the racemic cysteine solution. The aggregation of AgNPs was selectively induced by D-Cys, which allowed the precipitation of D-Cys with AgNPs (Fig. 4.4b), and left a net excess of the other enantiomer in the solution, thus resulting in enantio-separation. Similar to Fig. 4.4c, it was found that no significant UV–vis spectral shift (A_{520}/A_{400}) was detectable in the resultant supernatant (Fig. 4.4d).

We further exploited the effects of the UTP analogs (ATP, GTP, and CTP) on the stabilizing AgNPs toward the addition of L- and D-Cys. In our optimized experiment, the best candidate for chiral discrimination of Cys was the UTP-capped AgNPs, which showed an appreciable higher A_{520}/A_{400} discrepancy toward L-Cys and D-Cys.

The rational design and construction of optical active ligand-capped nanopar-ticles for enantioseparation and chiral detection is of intense current interest.

In this study, we have developed a novel nanoparticle-based biosensing platform using UTP-capped AgNPs as probe element, and demonstrated its feasibility in the application of colorimetric recognition of enantiomers based on absorption chemistry. This novel AgNPs-based probe design offers many advantages, including simplicity of preparation and manipulation compared with other methods that employ specific strategies. More importantly, our present sensor can achieve the goal of translating an enantioselective molecular recognition event into an appreciable color change via the aggregation of nanoparticles. This is the first application of nucleotide-capped AgNPs based biosensing platform for chiral recognition, and opens up new opportunities for the design of more novel enantiosensing strategies and enantiospecific adsorbents, and expansion of its application in different fields.

4.4 Colorimetric Assay for Phthalates Using UTP-Modified Gold Nanoparticles Cross-Linked by Copper (II)

Phthalates are chemical compounds primarily used as plasticizers which were added to plastics to increase their flexibility, transparency, durability, and longevity. In view of health concerns [51], phthalates are being phased out of many products in the United States, Canada, and European Union. Mono-ethyl-hexylphthalate, a metabolite of di (2-ethyl-hexyl) phthalate (DEHP), has been found to interact with all three peroxisome proliferator-activated receptors (PPARs), the members of the nuclear receptor superfamily. Phthalates belong to a class of metabolic disruptors that activate the roles of PPARs in lipid and carbohydrate metabolism [52]. Some phthalates, such as DEHP, dimethyl phthalate (DMP), and di (n-octyl) phthalate (DNOP), can affect hormone balance in the human body. In April 2011, a scandal was reported that phthalates, such as DEHP, and diisononyl phthalate (DINP), are illegally added to certain foods and beverages manufactured in Taiwan. DEHP and DINP have legitimate uses as plasticizers in food contact-packaging material, but they are not permitted as food additives. The toxicity of DEHP was about 30 times that of melamine, and this scandal was no less formidable than the melamine-tainted dairy products scandal that was detected in China in 2008. Long-time ingestion of DEHP at levels above the safety limit (1.5 ppm in Taiwan and Hong Kong) can affect hormone balance in the human body, thus resulting in the confusion of baby gender, the decrease of male reproductive ability, and female precocious puberty [53, 54]. To be concerned, the scope of phthalate-tainted samples has still been reported to be expanded from the food industry to the field of health-care products and medicines, which has seriously jeopardized the public security and social stability. Therefore, it is of paramount importance and urgency to develop a reliable and highly sensitive sensor that can provide on-site and real-time detection of phthalates in food products and medicines. However, current analytical techniques for phthalates in food products, such as gas chromatography/mass spectroscopy (GC/MS), all require expensive and complicated instruments, making on-site and real-time phthalate sensing difficult.

Interaction of nucleotides with certain metal ions has been investigated by many groups [55–57]. Some researchers have revealed that Cu^{2+} can interact with uridine and its analogues [58]. Moreover, Cu^{2+} can form coordination complexes with various ligands. Recently, Cu^{2+}-phthalate complexes have been reported to be constructed on the basis of the interaction between Cu^{2+} and phthalates and can be potent in vitro antitumor agents with 'self-activating' metallo-nuclease and DNA-binding properties [59].

Enlightened by the above facts, we think that a cross-network can be formed between UTP and phthalates in the presence of Cu^{2+} acting as a cross-linker based on their interactions, and furthermore, this appealing event of cross-network forming can be integrated with the unique properties of gold nanoparticles to develop a novel colorimetric assay for phthalates. In this section, we present a simple and reliable colorimetric method for the quantitative determination of phthalates in aqueous solutions utilizing UTP-modified gold nanoparticles (U-AuNPs) as a color indicator and Cu^{2+} as a cross-linker [60]. The aggregation of U-AuNPs can be induced by phthalates in the presence of Cu^{2+} acting as a cross-linker, coupled with the color of U-AuNPs that changed from initial red to purple. The as-prepared U-AuNPs can be used as a selective indicator for phthalates.

The U-AuNPs were prepared by the surface modification of AuNPs with UTP. A TEM image shows that UTP-AuNPs are highly dispersed in aqueous solution and the average size is about 15 nm (Fig. 4.5c, image a). The as-prepared U-AuNPs were red in color and showed a characteristic absorption peak at 520 nm, which was ascribed to the surface plasmon resonance of the AuNPs. The U-AuNPs solution can be highly stabilized against aggregation due to the negative capping agent's electrostatic repulsion between AuNPs. In the presence of 10 g/L phthalates dissolved in ethanol, the U-AuNPs can be also highly stabilized against aggregation, which can be evidenced by the profile of the absorption spectrum and the TEM image (Fig. 4.5). When the U-AuNPs were challenged to Cu^{2+} with varied concentrations, the absorption ratio ($A_{610}/_{520}$) of U-AuNPs was sensitive to the concentrations of Cu^{2+}. Moreover, it is noteworthy that in the presence of 0.4 μM Cu^{2+}, the U-AuNPs can be highly stabilized against aggregation, however, in the simultaneous presence of 0.4 μM Cu^{2+} and 10 g/L phthalates, significant absorption changes of U-AuNPs can be detected (an obvious decrease in the plasmon absorption at 520 nm and a strong increase in the surface plasmon band from 600 to 650 nm). The simultaneous presence of phthalates and Cu^{2+}, a cross-network between U-AuNPs and phthalates was formed, in which Cu^{2+} acted as a cross-linker to induce the aggregation of U-AuNPs and resulted in shift of the SPR absorption band of U-AuNPs to a longer wavelength, and consequently a color change from red to blue or purple. The phthalate-stimulated aggregation of U-AuNPs cross-linked by Cu^{2+} was also verified by the TEM image. In order to test the specificity of Cu^{2+} acting as a cross-linker for detection of phthalates, certain competing metal ions including 0.4 μM K^+, Ba^{2+}, Na^+, Mn^{2+}, Zn^{2+}, Fe^{3+},

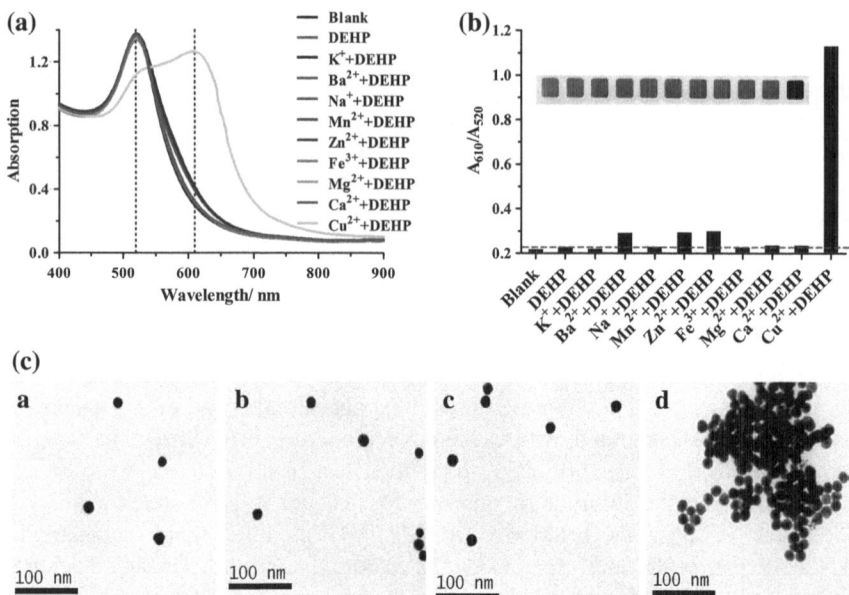

Fig. 4.5 **a** Typical absorption spectra of the U-AuNPs in the absence or presence of phthalates upon the simultaneous presence of various metal ions including 0.4 μM K^+, Ba^{2+}, Na^+, Mn^{2+}, Zn^{2+}, Fe^{3+}, Mg^{2+}, Ca^{2+}, and Cu^{2+}. **b** Bars represent the absorption ratio (A_{610}/A_{520}) of the U-AuNPs in the absence or presence of phthalates upon the simultaneous addition of various metal ions mentioned above. **c** TEM images of U-AuNPs (image a) and the U-AuNPs in the presence of 0.4 μM Cu^{2+} (image b), 10 g/L phthalates (image c), and 0.4 μM Cu^{2+} + 10 g/L phthalates (image d). Reprinted from Ref. [60] by permission of the Royal Society of Chemistry

Mg^{2+}, Ca^{2+} were examined under the same conditions as in the case of 0.4 μM Cu^{2+}. From the results, it was found that only the presence of Cu2+ can result in an obvious change in the absorption ratio (A_{610}/A_{520}) of U-AuNPs upon the addition of phthalates, accompanied with appreciable color switch, while there were nearly negligible changes in the absorption ratio (A_{610}/A_{520}) and color when in the presence of other metal stimulus.

We further exploited the effects of the UTP analogs (ATP, GTP, CTP, dATP, dGTP, dCTP, dTTP) on the stabilizing AuNPs toward the addition of Cu^{2+} and phthalates. In our optimized experiment, the best candidate for the detection of phthalates was the UTP-modified AuNPs, which showed an appreciable higher A_{610}/A_{520} for phthalates sensing. These experimental results demonstrated that copper (II) ion can form coordination complexes with UTP and phthalates.

Considering the appreciable changes in color and absorption properties of U-AuNPs toward phthalates, the potential of developing a novel colorimetric probe for the determination of phthalates was assessed. As a proof-of-concept experiment, the as-prepared U-AuNPs were challenged to DEHP, the most widely used phthalates, in the presence of Cu^{2+}. The kinetic behaviors of absorption ratio (A_{610}/A_{520}) of U-AuNPs responding to DEHP in the presence of 0.4 μM Cu^{2+} was

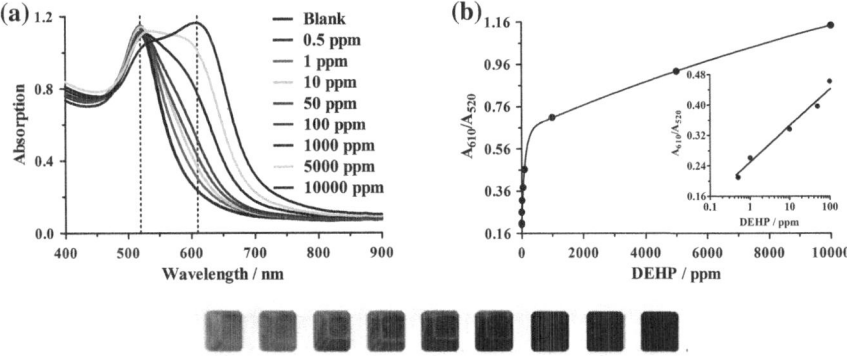

Fig. 4.6 **a** UV–vis absorbance of U-AuNPs in solution upon addition of varied concentration of DEHP (0, 0.5, 1, 10, 50, 100, 1000, 5000, and 10000 ppm) in the presence of 0.4 μM Cu^{2+} acting as a cross-linker. **b** Plot of absorption ratios A_{610}/A_{520} corresponding to the DEHP concentration in the range 0–10000 ppm in the presence of 0.4 μM Cu^{2+}. Inset: magnification of the plot of absorption ratios A_{610}/A_{520} corresponding to the DEHP concentration in the range 0–100 ppm in the presence of 0.4 μM Cu^{2+}. A Photo of U-AuNPs incubated with varied concentration of DEHP in the presence of 0.4 μM Cu^{2+} was listed. Reprinted from Ref. [60] by permission of the Royal Society of Chemistry

monitored. The phthalate-stimulated aggregation of U-AuNPs in the presence of Cu^{2+} was rapid, and the assay exhibited a nearly saturated signal within 5 min. Then, the different concentrations (0, 0.5, 1, 10, 50, 100, 1000, 5000, and 10000 ppm) of DEHP from one stock solution were added to the U-AuNPs probes in the presence of 0.4 μM Cu^{2+}. As presented in Fig. 4.6a, with the addition of an increasing concentration of DEHP to the suspension of U-AuNPs in the presence of 0.4 μM Cu^{2+}, an obvious decrease in the absorption peak at 520 nm and an increase in the absorption peak from 600 to 650 nm were clearly detected. These results were further confirmed by the color of the U-AuNPs which gradually changed from initially red to finally purple (image in Fig. 4.6). The sensitivity of the U-AuNPs probe for the DEHP was also investigated. From Fig. 4.6b, it can be seen that the absorption ratio (A_{610}/A_{520}) is sensitive to the concentration of DEHP. The limit of detection of DEHP using U-AuNPs probe was approximately 0.5 ppm, which was lower than the MAL (maximum allowable level) of DEHP (1.5 ppm) defined in Taiwan and Hong Kong. To the best of our knowledge, it is one of the most sensitive methods for the detection and analysis of phthalates, more importantly, this is the first application of AuNP-based probe for rapid and sensitive colorimetric assay for phthalates.

Furthermore, a variety of phthalates including DMP and DNOP and some competing stimulus including ethyl benzoate, phthalate acid, sodium oxalate, and sodium citrate were challenged to U-AuNPs probes in the presence of 0.4 μM Cu^{2+} under the same conditions as in the case of DEHP. The results demonstrated the excellent selectivity of this approach applied in phthalates detection (Fig. 4.7a, b). Therefore, a simple, sensitive, and reliable colorimetric probe for phthalates sensing

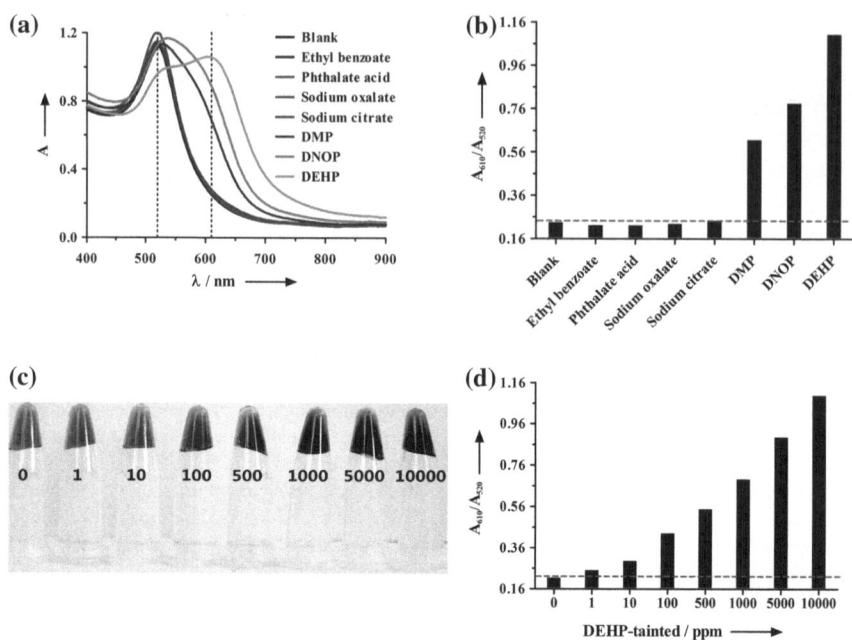

Fig. 4.7 **a** UV–vis absorbance of U-AuNPs in solution upon addition of various phthalates including 10 g/L of DMP, DNOP and DEHP and certain competing stimulus including 10 g/L ethyl benzoate, phthalate acid, sodium oxalate and sodium citrate. **b** Plot of absorption ratios A_{610}/A_{520} corresponding to analytes mentioned above. **c** Visual color change of the U-AuNPs sensor system without any addition and with the addition of raw samples tainted with 1, 10, 100, 500, 1000, 5000 and 10000 ppm DEHP. **d** The corresponding absorption ratio A_{610}/A_{520}. Reprinted from Ref. [60] by permission of the Royal Society of Chemistry

has been proved using U-AuNPs as a color indicator with Cu^{2+} acting as a cross-linker. To determine whether this colorimetric sensor could be applied to phthalates detection in real food and beverage samples, we first prepared a raw sample tainted with DEHP including tea drinks, carbonated drinks, juice drinks and vegetable protein drinks (detailed information in ref [60]). Then, a colorimetric assay for phthalate-tainted sample was conducted by the way described above. As shown in Fig. 4.7c, the sole addition of blank raw food and beverage to the U-AuNPs sensor solution did not lead to a distinguishable color change. However, when the samples with 1, 10, 100, 500, 1000, 5000 and 10000 ppm DEHP were added to the sensor solution, respectively, a red-to-purple color change could be clearly observed, coinciding with the increasing absorption ratio (A_{610}/A_{520}) demonstrated in Fig. 4.7d. Thus, the U-AuNPs system is a potential tool for rapid and sensitive assay of phthalates in food and beverage samples without complicated pretreatment and instrument.

Food safety has been one of the concerns in our daily life. There are so many scandals reporting that unessential substances are illegally added to certain foods and medicines resulting in human health poisoning, which has seriously jeopardized the public security and social stability. Therefore, it is of paramount importance and urgency to develop a reliable and highly sensitive sensor that can provide on-site and real-time detection of additives in food products and medicines.

In this work, we have developed a novel AuNP-based sensing probe using UTP-modified gold nanoparticles (U-AuNPs) as an indicator and demonstrated its feasibility in the application of colorimetric assay of phthalates in the presence of Cu^{2+} as a cross-linker based on absorption chemistry. This novel U-AuNPs probe offers many advantages, including simplicity of preparation and manipulation compared to other methods that employ specific strategies for phthalate detection. The aggregation of U-AuNPs was selectively induced by phthalates coupled with Cu^{2+}, which allowed the rapid colorimetric sensing of phthalates without complicated sample preparation and sophisticated instruments. It demonstrated superior sensitivity with a detection limit of phthalate of ca. 0.5 ppm, satisfactorily meeting the current MAL (1.5 ppm). To the best of our knowledge, it is one of the most sensitive methods for the detection and analysis of phthalates, more importantly, this is the first application of AuNP-based probe for colorimetric recognition of phthalates, and opens new opportunities for the design of more novel-sensing strategies and expansion of its application in different fields.

4.5 Multiplexed Analysis of Ag^+ and Hg^{2+} Using Oligonucleotide-Metal Nanoparticle Conjugates

Mercury and silver are two of the most hazardous metal pollutants, and they are widely distributed in ambient air, water, soil, and even food [61, 62]. Ag^+ ions are reported to inactivate sulfhydryl enzymes and accumulate in the body [63], and considered to be one of the heavy metal ions that can cause environmental pollution for water resources [64]. While long-term exposure to high levels of mercury can lead to a variety of adverse health effects, such as damage to the brain, nervous system, immune system, and many other organs [65, 66].

Hg^{2+} is reported to selectively coordinate thymine (T) bases and form stable T-Hg^{2+}-T complexes. Based on this concept, two strands of DNA, which are designed to be complementary except for a single T–T mismatch, are used to modify the surface of the metal nanoparticles. The resulting two types of DNA-functionalized AuNPs are selectively aggregated in the presence of Hg^{2+} based on T-Hg^{2+}-T coordination [67]. Like T-Hg^{2+}-T coordination chemistry, Ag^+ ion can also be exclusively captured by cytosine–cytosine (C–C) in DNA duplexes to form Ag^+-cytosine complexes. AgNPs functionalized with cytosine-(C)-rich oligonucleotide can selectively aggregate in the presence of Ag^+ based on the formation of C-Ag^+-C complexes [68], although these methods all show good sensitivity and selectivity to Hg^{2+} or Ag^+. However, the analysis of these two metal ions simultaneously utilizing AuNPs and AgNPs in a single step in homogeneous solution remains a challenge.

In this section, we unveil a simple method for simultaneously sensing Hg^{2+} and Ag^+ via a mixture of T-rich oligonucleotide-AgNPs (AgNPs-T-ssDNA) and C-rich oligonucleotide-AuNPs (AuNPs-C-ssDNA) conjugates [19]. Namely, the presence of Ag^+, the AuNPs-C-ssDNA color will be changed due to the aggregation of AuNPs in salt solution, from which Ag^+ can be recognized, while the presence of Hg^{2+} will change the color of AgNPs-T-ssDNA in salt solution. In cases where there is presence of these two metal ions, the preferred color of AuNPs-C-ssDNA and AgNPs-T-ssDNA both have been changed in salt solution. The color change can be identified with the naked eye or by UV–vis measurement.

According to previous reports [69], when HS-C-ssDNA is covalently labeled onto the AuNPs to produce the AuNPs-C-ssDNA complex, HS-C-ssDNA can enhance the stability of AuNPs against aggregation caused by salt. Solution of AuNPs-C-ssDNA appears red, and has a surface plasma resonance absorption peak at about 520 nm in the UV–vis spectra. However, such stability can still be obviously destroyed by the hybridization of HS-C-ssDNA tagged on the AuNPs with its complementary strand (Ag^+-stimulated forming of C-Ag^+-C complex). That is because the repulsive interactions from phosphate back-bones on ssDNA are greatly decreased by formation of the rigid double strands.

In this work, in the presence of different concentrations of Ag^+, the AuNPs-C-ssDNA/Ag^+ colloid aggregated easily in $NaNO_3$-MOPS buffer coupled with color change from red to purple (Fig. 4.8a, 60 nM, 600 nM and 3.0 μM) even dark blue with higher concentration of Ag^+ (Fig. 4.8a, 10 and 20 μM) within 1 min. Additionally, in the UV–vis spectra, plasmon peaks at 520 nm of AuNPs decreased and the ones at 650 nm increased with increasing concentration of Ag^+ (Fig. 4.8b). The reason for such results is that, in the presence of Ag^+, each Ag^+ ion combines two C-bases in the neighboring strands on the same AuNPs or on the other AuNPs, forming the double strand structure, so the stability of AuNPs was obviously destroyed as described above. In the optimized conditions, the concentration of Ag^+ of 60 nM can be being discriminated by naked eyes (Fig. 4.8a). The absorption ratio between 650 and 520 nm (A_{650}/A_{520}) was employed to quantitatively determine the concentration of Ag^+. As shown in Fig. 4.8c, the A_{650}/A_{520} ratio showed a sensitive function toward Ag^+ concentrations. A linear correlation was obtained between the A_{650}/A_{520} ratio and the concentration of Ag^+ ions from 0 to 200 nM ($R^2 = 0.9993$), and the limit of detection (LOD) of Ag^+ based on the three times of signal-to-noise level of the blank sample was estimated to be 1 nM. This detection limit was better than those of previous reports using fluorescence probe (10 nM) [70], chromogenic method (130 nM) [71], and DNAzyme biosensor (6.3 nM) [72]. A series of 8 independent measurements of 100 nM Ag^+ were used for estimating the precision, and the relative standard deviation (RSD) was calculated to be 3.2%, which showed that the proposed AuNP-C-ssDNA biosensor had good reproducibility.

To demonstrate the selectivity of colloidal solutions of AuNPs-C-ssDNA toward Ag^+, other competing metal ions including Zn^{2+}, Cu^{2+}, Co^{2+}, Ni^{2+}, Ca^{2+}, Mg^{2+}, Mn^{2+}, Fe^{3+}, Cd^{2+}, Pb^{2+}, Hg^{2+}, Al^{3+} were examined under the same conditions as Ag^+. As a result, in the presence of 20 μM of Ag^+, the color of the

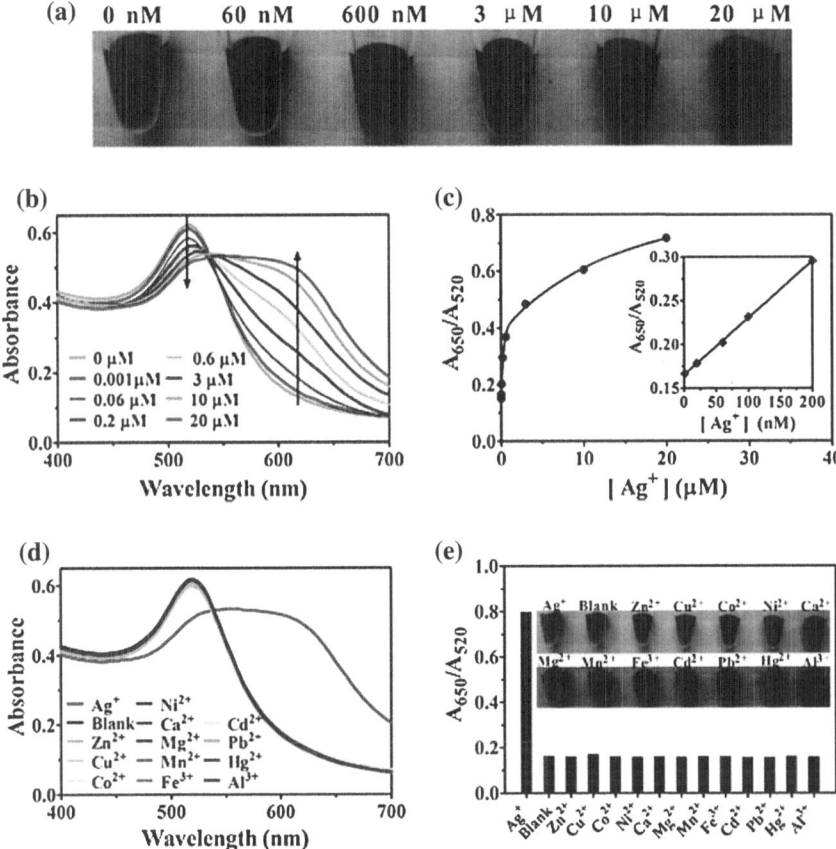

Fig. 4.8 a Digital photo of colorimetric AuNPs-C-ssDNA sensor response to different concentrations of Ag$^+$ in NaNO$_3$-MOPS buffer. **b** UV–vis absorption spectra response of AuNPs-C-ssDNA upon different concentrations of Ag$^+$ in NaNO$_3$-MOPs buffer. **c** Plot of the absorption ratio (A$_{650}$/A$_{520}$) of different concentrations of Ag$^+$ (from 0 nM to 20 μM). Inset: Plot of the absorption ratio (A$_{650}$/A$_{520}$) of Ag$^+$ from 1 nM to 200 nM. **d** UV–vis absorption spectra response of AuNPs-C-ssDNA upon 20 μM Ag$^+$ and 20 μM other competing metal ions in NaNO$_3$-MOPS buffer. **e** The absorption ratio (A$_{650}$/A$_{520}$) in the presence of 20 μM Ag$^+$ and 20 μM other competing metal ions. Inset: Color change of AuNPs-C-ssDNA sensing system treated with 20 μM Ag$^+$ and 20 μM other competing metal ions. Reprinted from Ref. [19] by permission of the Royal Society of Chemistry

colloidal solutions of AuNPs-C-ssDNA/Ag$^+$ quickly turned from red to dark blue (because each Ag$^+$ ion combined two C-bases in the neighboring strands forming the double strand structure). But when Ag$^+$ were replaced by 20 μM of other metal ions, the color of the solutions AuNPs-C-ssDNA/Metal ions remained almost similar to the blank (Fig. 4.8e, inset). Therefore, the competing metal ions mentioned above cannot bring interferences to Ag$^+$ detection in this work.

Like AuNPs-C-ssDNA colloidal solutions, colloidal solutions of AgNPs-T-ssDNA can also provide high stability against salt, and such stability can still be obviously destroyed by the hybridization of SH-T-ssDNA tagged on the AgNPs with its complementary strand (Hg^{2+}-stimulated forming of T-Hg^{2+}-T complex), leading to easier aggregations of AgNPs colloid and display different colors. Fig. 4.9 shows UV–vis absorption spectra and the corresponding photographs of AgNPs-T-ssDNA colloidal solutions treated with 0–20 μM Hg^{2+} in $NaNO_3$-MOPS buffer. After the different concentrations of Hg^{2+} were added into AgNPs-T-ssDNA colloidal solutions, each Hg^{2+} ion combined two T-bases in the neighboring strands forming the double strand structure, the AgNPs-T-ssDNA/Hg^{2+} colloid aggregated easily in salt solution, and displayed different colors. Here, the color change was also highly dependent on the concentrations of Hg^{2+} (Fig. 4.9a). In the UV–vis absorption spectra, with the increase of Hg^{2+} concentration, the plasmon peaks at 390 nm gradually decreased and meanwhile the shoulder absorption band at 530 nm gradually increased. It should be noted that in this system, when the concentration of Hg^{2+} in AgNPs-T-ssDNA solutions increased, changes in UV–vis spectra of AgNPs were mainly manifested in the decline of plasmon peaks at 390 nm, meanwhile the increase of shoulder absorption band at 530 nm was not too high, and it was only evident with the presence of relatively higher concentration of Hg^{2+} (Fig. 4.9b). In the optimized conditions, when the concentration of Hg^{2+} was as low as 100 nM, a slight color change of AgNPs-T-ssDNA colloidal solution can be discriminated by naked eyes (Fig. 4.9a). The absorption ratio A_{530}/A_{390} was employed to quantitatively determine the concentration of Hg^{2+}. A linear correlation was obtained between the A_{530}/A_{390} and the concentration of Hg^{2+} ions from 0 to 200 nM ($R^2 = 0.9962$) with the lowest detectable concentration of 1 nM based on the three times of signal-to-noise level of the blank sample (Fig. 4.9c inset). This sensitivity was better than those of previous reports using fluorescence (40 nM) [73], sensitive electrochemical method (10 nM) [74], and had the same performance as surfactant-templated mesoporous silica based method (2 nM) [75]. The RSD was 4.3% corresponding to 100 nM Hg^{2+} (n = 8). It showed that the proposed AgNP-T-ssDNA biosensor had good reproducibility.

The selectivity of AgNPs-T-ssDNA colloidal solution toward Hg^{2+} (Fig. 4.9d, e) was similar with that of the selectivity of AuNPs-C-ssDNA colloidal solution toward Ag^+ as discussed above. Experimental results given in Fig. 4.9d, e indicate that through color observation and UV–vis absorption spectra, the competing metal ions (Zn^{2+}, Cu^{2+}, Co^{2+}, Ni^{2+}, Ca^{2+}, Mg^{2+}, Mn^{2+}, Fe^{3+}, Cd^{2+}, Pb^{2+}, Ag^+, Al^{3+}) cannot bring interferences to Hg^{2+} detection in this work.

A mixture of 11.5 nM AuNPs-C-ssDNA and 8.4 nM AgNPs-T-ssDNA colloidal solutions with equal volume was prepared and treated with different concentrations (0–20 μM) of Ag^+ or Hg^{2+} or different concentrations (0–20 μM) of a mixture of Ag^+ and Hg^{2+}, and finally a sufficient amount of salt (0.05 M $NaNO_3$ as final concentration) was added before discriminated by naked eyes or UV–vis measurement (Fig. 4.10). In the presence of Ag^+ (0–20 μM) and absence of Hg^{2+} in the mixture of AuNPs-C-ssDNA and AgNPs-T-ssDNA solutions, the plasmon

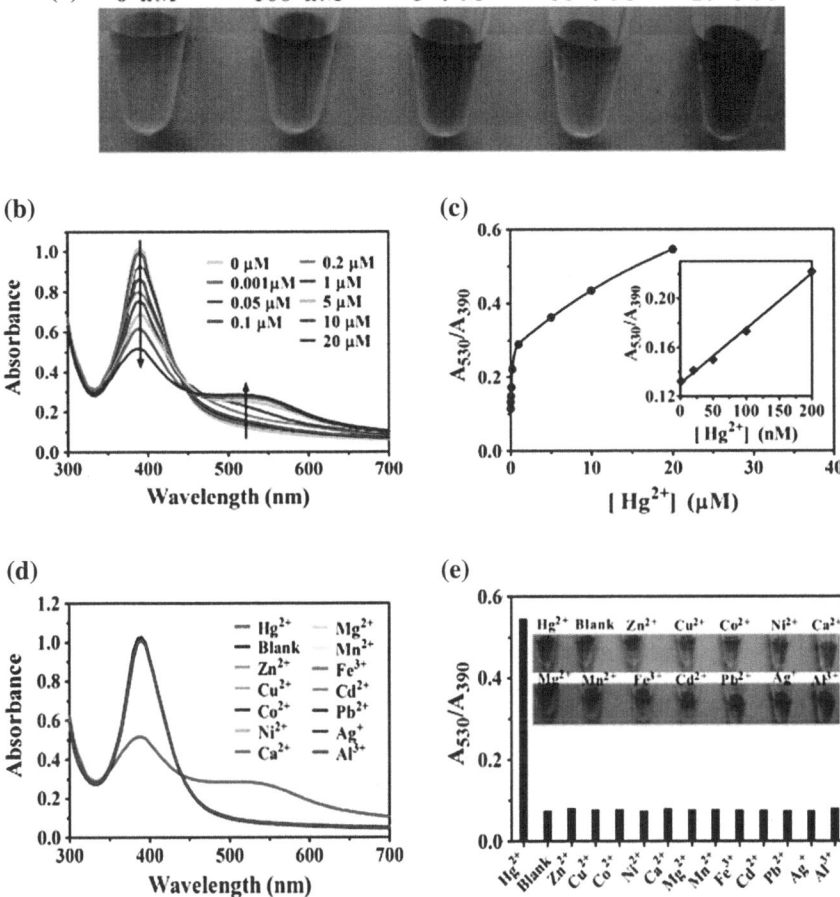

Fig. 4.9 a Digital photo of colorimetric AgNPs-T-ssDNA sensor response to different concentrations of Hg^{2+} in NaNO$_3$-MOPS buffer. **b** UV–vis absorption spectra response of AgNPs-T-ssDNA upon different concentrations of Hg^{2+} in NaNO$_3$-MOPS buffer. **c** Plot of the absorption ratio (A$_{530}$/A$_{390}$) of different concentrations of Hg^{2+} (from 0 nM to 20 μM). Inset: Plot of the absorption ratio (A$_{530}$/A$_{390}$) of Hg^{2+} from 1 nM to 200 nM. **d** UV–vis absorption spectra response of AgNPs-T-ssDNA upon 20 μM Hg^{2+} and 20 μM other competing metal ions in NaNO$_3$-MOPS buffer. **e** The absorption ratio (A$_{530}$/A$_{390}$) in the presence of 20 μM Hg^{2+} and 20 μM of other competing metal ions in NaNO$_3$-MOPS buffer. Inset: Color change of AgNPs-T-ssDNA sensing system treated with 20 μM Hg^{2+} and 20 μM other metal ions. Reprinted from Ref. [19] by permission of the Royal Society of Chemistry

peaks at 520 nm of AuNPs gradually decreased and the shoulder absorption band at 650 nm increased with increasing concentrations of Ag$^+$, while the plasmon peaks at 390 nm of AgNPs showed no change before and after the addition of Ag$^+$. The results described above can be interpreted by the formation of double strand structure from single strands HS-C-ssDNA on the AuNPs through forming

C-Ag$^+$-C base pairs in DNA duplexes. These suggested that the extinction display of AuNPs did not affect the plasmon peaks at 390 nm of AgNPs in the system. Conversely, when Hg^{2+} was added into the mixture of AuNPs-C-ssDNA and AgNPs-T-ssDNA solutions, along with the increase of Hg^{2+} concentration, the plasmon peaks at 390 nm of AgNPs gradually decreased, while the plasmon peaks at 520 nm and the shoulder absorption band at 650 nm of AuNPs slightly increased with the concentration of Hg^{2+} increasing from 1 μM to 20 μM. In this case, the change of the plasmon peaks at 390 nm of AgNPs can be interpreted by the formation of double strand structure from single strands HS-T-ssDNA on the AgNPs through forming T-Hg^{2+}-T base pairs. And in the presence of both Hg^{2+} and Ag$^+$, the plasmon peaks at 390 nm of AgNPs and the plasmon peaks at 520 nm of AuNPs gradually decreased with increasing concentrations of Hg^{2+} and Ag$^+$, moreover, the shoulder absorption band at 650 nm increased (Fig. 4.10b). To assess the impact of Hg^{2+} concentration on the results of quantitative determination of Ag$^+$ concentration in this method, a series of combinations of metal ion solutions (5 nM Ag$^+$ + 5 μM Hg^{2+}), (5 nM Ag$^+$ + 10 μM Hg^{2+}), or (5 nM Ag$^+$ + 20 μM Hg^{2+}) were added into a mixture of AuNPs-C-ssDNA and AgNPs-T-ssDNA colloidal solutions. As a result, there were changes of the plasmon peaks at 520 nm and absorption at 650 nm of AuNPs, but the absorption ratio at A_{650}/A_{520} did not change with different concentrations of Hg^{2+}. This result showed that the presence of high concentrations of Hg^{2+} did not affect the results of quantitative determination of Ag$^+$ in this system. Moreover, the increase of the plasmon peaks at 520 nm and absorption at 650 nm of AuNPs in the presence of high concentrations of Hg^{2+} was not due to direct effects of Hg^{2+} on AuNPs, but due to impact of the color change of AgNPs (from pale yellow to dark purple) in the presence of high concentration of Hg^{2+}.

In this part, the absorption at 390 nm was employed to quantitatively determine the concentration of Hg^{2+}. A linear correlation was obtained between the absorption (A_{390}) and the concentration of Hg^{2+} from 0 to 100 nM ($R^2 = 0.9907$), and a detection limit of 5 nM was obtained using the method of 3σ (Fig. 4.10c, inset). The absorption ratio A_{650}/A_{520} was employed to quantitatively determine the concentration of Ag$^+$. A linear correlation was obtained between the absorption ratio (A_{650}/A_{520}) and the concentration of Hg^{2+} from 0 to 200 nM ($R^2 = 0.9913$), and a detection limit of 5 nM was obtained using the method of 3σ (Fig. 4.10d, inset). The selectivity of the system of AgNPs-T-ssDNA and AuNPs-C-ssDNA toward Ag$^+$ and Hg^{2+} was also tested; the results shown in Fig. 4.10e indicated that the competing metal ions (Zn^{2+}, Cu^{2+}, Co^{2+}, Ni^{2+}, Ca^{2+}, Mg^{2+}, Mn^{2+}, Fe^{3+}, Cd^{2+}, Pb^{2+}, Al^{3+}) bring little interference to Ag$^+$ and Hg^{2+} detection in this work.

To test the potential of the proposed method for simultaneous analysis of Hg^{2+} and Ag$^+$ in environmental samples, a water sample was collected from the Qin Chun River in the campus of the East China University of Science and Technology. The sample was first filtered through a 0.22 μm membrane to remove soil and other particles, and then tested by the proposed technique. As a result, no difference in UV–vis absorption spectra of AuNPs-C-ssDNA/AgNPs-T-ssDNA sensing system occurred in the measured sample of river water and blank,

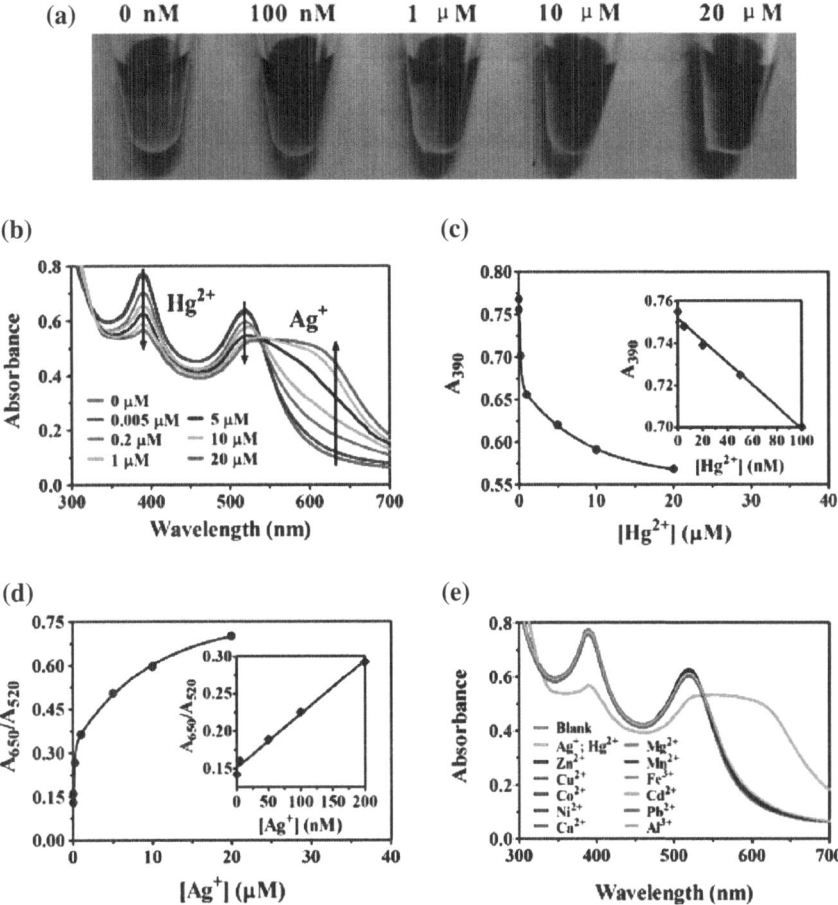

Fig. 4.10 **a** Color change of AuNPs-C-ssDNA/AgNPs-T-ssDNA sensing system treated with different concentrations of Ag^+ and Hg^{2+} mixture. **b** UV–vis absorption spectra of AuNPs-C-ssDNA/AgNPs-T-ssDNA sensing system treated with different concentrations of Ag^+ and Hg^{2+} mixture. **c** Plot of the absorption (A_{390}) of different concentrations of Hg^{2+} (from 0 nM to 20 μM). Inset: Plot of the absorption (A_{390}) of Hg^{2+} from 1 nM to 100 nM. **d** Plot of the absorption ratio (A_{650}/A_{520}) of different concentrations of Ag^+ (from 0 nM to 20 μM). Inset: Plot of the absorption ratio (A_{650}/A_{520}) of Ag^+ from 1 to 200 nM. Reprinted from Ref. [19] by permission of the Royal Society of Chemistry

indicating that Hg^{2+} and Ag^+ were not found in this water sample. The result was verified using our recently reported graphene-based fluorescence biosensor (data not shown) [76]. Then, the water samples were spiked with mixtures of Hg^{2+} and Ag^+ at different concentration levels, the recoveries of 5, 50, and 100 nM Hg^{2+} were $110.2 \pm 6.8\%$, $95.6 \pm 2.2\%$ and $82.8 \pm 2.6\%$, and the recoveries of 5, 50, and 100 nM Ag^+ were $106.5 \pm 5.6\%$, $96.3 \pm 2.6\%$ and $88.2 \pm 1.8\%$, respectively. Therefore, the proposed method is attractive for monitoring low levels of Hg^{2+} and Ag^+ in water samples.

In conclusion, this work revealed a novel assay method for highly sensitive and selective detection of Hg^{2+} and Ag^+ by using a mixture of AuNPs-C-ssDNA/AgNPs-T-ssDNA solution. This method allows for selective determination of Hg^{2+} and Ag^+, and lowest concentration of Hg^{2+} and Ag^+ mixture detected by the UV–vis spectra measurement was 5 nM, which is comparable with the most sensitive methods that have been previously reported for detecting Hg^{2+} and Ag^+. In addition, the use of AuNPs-C-ssDNA-sensing system or AgNPs-T-ssDNA sensing system separately to detect Ag^+ or Hg^{2+} also achieved a lower detection limit of 1 nM. This sensing system was successfully applied to the sensitive and selective detection of Hg^{2+} and Ag^+ in the water sample. We believe that the present approach holds great potential for monitoring Hg^{2+} and Ag^+ in environmental samples.

References

1. Fan CH, Wang LH, Li J, Song SP, Li D (2009) Biomolecular sensing via coupling DNA-based recognition with gold nanoparticles. J Phys D Appl Phys 42(20):203001. doi:10.1088/0022-3727/42/20/203001
2. Iqbal SS, Mayo MW, Bruno JG, Bronk BV, Batt CA, Chambers JP (2000) A review of molecular recognition technologies for detection of biological threat agents. Biosens Bioelectron 15(11–12):549–578. doi:10.1016/S0956-5663(00)00108-1
3. Giljohann DA, Mirkin CA (2009) Drivers of biodiagnostic development. Nature 462(7272):461–464. doi:10.1038/nature08605
4. Elghanian R, Storhoff JJ, Mucic RC, Letsinger RL, Mirkin CA (1997) Selective colorimetric detection of polynucleotides based on the distance-dependent optical properties of gold nanoparticles. Science 277(5329):1078–1081. doi:10.1126/science.277.5329.1078
5. Thompson DG, Enright A, Faulds K, Smith WE, Graham D (2008) Ultrasensitive DNA detection using oligonucleotide-silver nanoparticle conjugates. Anal Chem 80(8):2805–2810. doi:10.1021/ac702403w
6. Han MS, Lytton-Jean AK, Oh BK, Heo J, Mirkin CA (2006) Colorimetric screening of DNA-binding molecules with gold nanoparticle probes. Angew Chem Int Ed Engl 45(11):1807–1810. doi:10.1002/anie.200504277
7. Zhao W, Chiuman W, Lam JC, McManus SA, Chen W, Cui Y, Pelton R, Brook MA, Li Y (2008) DNA aptamer folding on gold nanoparticles: from colloid chemistry to biosensors. J Am Chem Soc 130(11):3610–3618. doi:10.1021/ja710241b
8. Li F, Zhang J, Cao X, Wang L, Li D, Song S, Ye B, Fan C (2009) Adenosine detection by using gold nanoparticles and designed aptamer sequences. Analyst 134(7):1355–1360. doi:10.1039/b900900k
9. Zhang J, Wang L, Pan D, Song S, Boey FY, Zhang H, Fan C (2008) Visual cocaine detection with gold nanoparticles and rationally engineered aptamer structures. Small 4(8):1196–1200. doi:10.1002/smll.200800057
10. Liu J, Lu Y (2005) Fast colorimetric sensing of adenosine and cocaine based on a general sensor design involving aptamers and nanoparticles. Angew Chem Int Ed Engl 45(1):90–94. doi:10.1002/anie.200502589
11. Thaxton CS, Elghanian R, Thomas AD, Stoeva SI, Lee JS, Smith ND, Schaeffer AJ, Klocker H, Horninger W, Bartsch G, Mirkin CA (2009) Nanoparticle-based bio-barcode assay

redefines "undetectable" PSA and biochemical recurrence after radical prostatectomy. Proc Natl Acad Sci U S A 106(44):18437–18442. doi:10.1073/pnas.0904719106

12. Tessier PM, Jinkoji J, Cheng YC, Prentice JL, Lenhoff AM (2008) Self-interaction nanoparticle spectroscopy: a nanoparticle-based protein interaction assay. J Am Chem Soc 130(10):3106–3112. doi:10.1021/ja077624q

13. Goto-Inoue N, Hayasaka T, Zaima N, Kashiwagi Y, Yamamoto M, Nakamoto M, Setou M (2010) The detection of glycosphingolipids in brain tissue sections by imaging mass spectrometry using gold nanoparticles. J Am Soc Mass Spectrom 21(11):1940–1943. doi:10.1016/j.jasms.2010.08.002

14. Scampicchio M, Arecchi A, Mannino S (2009) Optical nanoprobes based on gold nanoparticles for sugar sensing. Nanotechnology 20(13):135501. doi:10.1088/0957-4484/20/13/135501

15. Zhang L, Yao Y, Shan J, Li H (2011) Lead (II) ion detection in surface water with pM sensitivity using aza-crown-ether-modified silver nanoparticles via dynamic light scattering. Nanotechnology 22(27):275504. doi:10.1088/0957-4484/22/27/275504

16. Huy GD, Zhang M, Zuo P, Ye BC (2011) Multiplexed analysis of silver(I) and mercury(II) ions using oligonucletide-metal nanoparticle conjugates. Analyst 136(16):3289–3294. doi:10.1039/c1an15373k

17. Lin CY, Yu CJ, Lin YH, Tseng WL (2010) Colorimetric sensing of silver(I) and mercury(II) ions based on an assembly of Tween 20-stabilized gold nanoparticles. Anal Chem 82(16):6830–6837. doi:10.1021/ac1007909

18. Daniel WL, Han MS, Lee JS, Mirkin CA (2009) Colorimetric nitrite and nitrate detection with gold nanoparticle probes and kinetic end points. J Am Chem Soc 131(18):6362–6363. doi:10.1021/ja901609k

19. Ye BC, Huy GD, Zhang M, Zuo P (2011) Multiplexed analysis of silver(I) and mercury(II) ions using oligonucletide-metal nanoparticle conjugates. Analyst 136(16):3289–3294. doi:10.1039/C1an15373k

20. Yu AB, Jiang XC (2008) Silver nanoplates: A highly sensitive material toward inorganic anions. Langmuir 24(8):4300–4309. doi:10.1021/La7032252

21. Pal T, Ghosh SK (2007) Interparticle coupling effect on the surface plasmon resonance of gold nanoparticles: From theory to applications. Chem Rev 107(11):4797–4862. doi:10.1021/Cr0680282

22. Pal T, Ghosh SK, Kundu S, Mandal M (2002) Silver and gold nanocluster catalyzed reduction of methylene blue by arsine in a micellar medium. Langmuir 18(23):8756–8760. doi:10.1021/La0201974

23. Pal T, Ghosh SK, Pal A, Nath S, Kundu S, Panigrahi S (2005) Dimerization of eosin on nanostructured gold surfaces: Size regime dependence of the small metallic particles. Chem Phys Lett 412(1–3):5–11. doi:10.1016/j.cplett.2005.06.059

24. Sohn Y, Pradhan D, Radi A, Leung KT (2009) Interfacial electronic structure of gold nanoparticles on Si(100): alloying versus quantum size effects. Langmuir 25(16):9557–9563. doi:10.1021/la900828v

25. Zhou HS, Honma II, Komiyama H, Haus JW (1994) Controlled synthesis and quantum-size effect in gold-coated nanoparticles. Phys Rev B: Condens Matter 50(16):12052–12056. doi:10.1103/PhysRevB.50.12052

26. Chhabra R, Sharma J, Wang H, Zou S, Lin S, Yan H, Lindsay S, Liu Y (2009) Distance-dependent interactions between gold nanoparticles and fluorescent molecules with DNA as tunable spacers. Nanotechnology 20(48):485201. doi:10.1088/0957-4484/20/48/485201

27. Sih BC, Wolf MO (2006) Dielectric medium effects on collective surface plasmon coupling interactions in oligothiophene-linked gold nanoparticles. J Phys Chem B 110(45):22298–22301. doi:10.1021/jp065213a

28. Ivanov AR, Nazimov IV, Baratova L (2000) Determination of biologically active low-molecular-mass thiols in human blood I. Fast qualitative and quantitative, gradient and isocratic reversed-phase high-performance liquid chromatography with photometric and

fluorescence detection. J Chromatogr A 895(1–2):157–166. doi:10.1016/S0021-9673(00)00713-5

29. Jacobsen DW (1998) Homocysteine and vitamins in cardiovascular disease. Clin Chem 44(8 Pt 2):1833–1843

30. Ames BN, Liu JK, Yeo HC, Overvik-Douki E, Hagen T, Doniger SJ, Chu DW, Brooks GA (2000) Chronically and acutely exercised rats: biomarkers of oxidative stress and endogenous antioxidants. J Appl Physiol 89(1):21–28

31. Hartleb J, Arndt R (2001) Cysteine and indole derivatives as markers for malignant melanoma. J Chromatogr B Biomed Sci Appl 764(1–2):409–443. doi:10.1016/S0378-4347(01)00278-X

32. Lee JS, Ulmann PA, Han MS, Mirkin CA (2008) A DNA-gold nanoparticle-based colorimetric competition assay for the detection of cysteine. Nano Lett 8(2):529–533. doi:10.1021/nl0727563

33. Li L, Li B (2009) Sensitive and selective detection of cysteine using gold nanoparticles as colorimetric probes. Analyst 134(7):1361–1365. doi:10.1039/b819842j

34. Lee JS, Lytton-Jean AK, Hurst SJ, Mirkin CA (2007) Silver nanoparticle-oligonucleotide conjugates based on DNA with triple cyclic disulfide moieties. Nano Lett 7(7):2112–2115. doi:10.1021/nl071108g

35. Kimura-Suda H, Petrovykh DY, Tarlov MJ, Whitman LJ (2003) Base-dependent competitive adsorption of single-stranded DNA on gold. J Am Chem Soc 125(30):9014–9015. doi:10.1021/ja035756n

36. Ostblom M, Liedberg B, Demers LM, Mirkin CA (2005) On the structure and desorption dynamics of DNA bases adsorbed on gold: a temperature-programmed study. J Phys Chem B 109(31):15150–15160. doi:10.1021/jp051617b

37. Breier A, Ziegelhoffer A, Famulsky K, Michalak M, Slezak J (1996) Is cysteine residue important in FITC-sensitive ATP-binding site of P-type ATPases? A commentary to the state of the art. Mol Cell Biochem 160–161:89–93. doi:10.1007/BF00240036

38. Shang L, Yin J, Li J, Jin L, Dong S (2009) Gold nanoparticle-based near-infrared fluorescent detection of biological thiols in human plasma. Biosens Bioelectron 25(2):269–274. doi:10.1016/j.bios.2009.06.021

39. Chang HT, Lee KH, Chen SJ, Jeng JY, Cheng YC, Shiea JT (2007) Fluorescence and interactions with thiol compounds of Nile Red-adsorbed gold nanoparticles. J Colloid Interf Sci 307(2):340–348. doi:10.1016/j.jcis.2006.12.013

40. Kusmierek K, Glowacki R, Bald E (2006) Analysis of urine for cysteine, cysteinylglycine, and homocysteine by high-performance liquid chromatography. Anal Bioanal Chem 385(5):855–860. doi:10.1007/s00216-006-0454-x

41. Fernig DG, Doty RC, Tshikhudo TR, Brust M (2005) Extremely stable water-soluble Ag nanoparticles. Chem Mater 17(18):4630–4635. doi:10.1021/Cm0508017

42. Janaky R, Varga V, Hermann A, Saransaari P, Oja SS (2000) Mechanisms of L-cysteine neurotoxicity. Neurochem Res 25(9–10):1397–1405. doi:10.1023/A:1007616817499

43. Plateau P, Soutourina J, Blanquet S (2001) Role of D-cysteine desulfhydrase in the adaptation of Escherichia coli to D-cysteine. J Biol Chem 276(44):40864–40872. doi:10.1074/jbc.M102375200

44. Gautier C, Burgi T (2009) Chiral gold nanoparticles. ChemPhysChem 10(3):483–492. doi:10.1002/cphc.200800709

45. Noguez C, Garzon IL (2009) Optically active metal nanoparticles. Chem Soc Rev 38(3):757–771. doi:10.1039/b800404h

46. Shukla N, Bartel MA, Gellman AJ (2010) Enantioselective separation on chiral Au nanoparticles. J Am Chem Soc 132(25):8575–8580. doi:10.1021/ja908219h

47. Wang Y, Yin X, Shi M, Li W, Zhang L, Kong J (2006) Probing chiral amino acids at sub-picomolar level based on bovine serum albumin enantioselective films coupled with silver-enhanced gold nanoparticles. Talanta 69(5):1240–1245. doi:10.1016/j.talanta.2005.12.060

48. Ye BC, Zhang M (2011) Colorimetric Chiral Recognition of Enantiomers Using the Nucleotide-Capped Silver Nanoparticles. Anal Chem 83(5):1504–1509. doi:10.1021/Ac102922f

49. Lim II, Mott D, Engelhard MH, Pan Y, Kamodia S, Luo J, Njoki PN, Zhou S, Wang L, Zhong CJ (2009) Interparticle chiral recognition of enantiomers: a nanoparticle-based regulation strategy. Anal Chem 81(2):689–698. doi:10.1021/ac802119p

50. Oja SS, Janaky R, Varga V, Hermann A, Saransaari P (2000) Mechanisms of L-cysteine neurotoxicity. Neurochem Res 25(9–10):1397–1405. doi:10.1023/A:1007616817499

51. Lopez-Carrillo L, Hernandez-Ramirez RU, Calafat AM, Torres-Sanchez L, Galvan-Portillo M, Needham LL, Ruiz-Ramos R, Cebrian ME (2010) Exposure to phthalates and breast cancer risk in northern Mexico. Environ Health Perspect 118(4):539–544. doi:10.1289/ehp.0901091

52. Desvergne B, Feige JN, Casals-Casas C (2009) PPAR-mediated activity of phthalates: A link to the obesity epidemic? Mol Cell Endocrinol 304(1–2):43–48. doi:10.1016/j.mce.2009.02.017

53. Botelho GG, Golin M, Bufalo AC, Morais RN, Dalsenter PR, Martino-Andrade AJ (2009) Reproductive effects of di(2-ethylhexyl)phthalate in immature male rats and its relation to cholesterol, testosterone, and thyroxin levels. Arch Environ Contam Toxicol 57(4):777–784. doi:10.1007/s00244-009-9317-8

54. Zheng SJ, Tian HJ, Cao J, Gao YQ (2010) Exposure to di(n-butyl)phthalate and benzo(a)pyrene alters IL-1beta secretion and subset expression of testicular macrophages, resulting in decreased testosterone production in rats. Toxicol Appl Pharmacol 248(1):28–37. doi:10.1016/j.taap.2010.07.008

55. Gasowska A, Jastrzab R, Lomozik L (2007) Specific type of interactions in the quaternary system of Cu(II), adenosine 5′-triphosphate, 1,11-diamino-4,8-diazaundecane and uridine. J Inorg Biochem 101(10):1362–1369. doi:10.1016/j.jinorgbio.2007.05.009

56. Knobloch B, Mucha A, Operschall BP, Sigel H, Jezowska-Bojczuk M, Kozlowski H, Sigel RK (2011) Stability and structure of mixed-ligand metal ion complexes that contain Ni2 + , Cu2 + , or Zn2 + , and Histamine, as well as adenosine 5′-triphosphate (ATP4-) or uridine 5′-triphosphate (UTP(4-): an intricate network of equilibria. C. Chemistry 17(19):5393–5403. doi:10.1002/chem.201001931

57. Tu AT, Friederich CG (1968) Interaction of copper ion with guanosine and related compounds. Biochemistry 7(12):4367–4372. doi:10.1021/bi00852a032

58. Lomozik L, Jastrzab R (2003) Copper(II) complexes with uridine, uridine 5 '-monophosphate, spermidine, or spermine in aqueous solution. J Inorg Biochem 93(3–4):132–140. doi:10.1016/S0162-0134(02)00567-6

59. Kellett A, O'Connor M, McCann M, McNamara M, Lynch P, Rosair G, McKee V, Creaven B, Walsh M, McClean S, Foltyn A, O'Shea D, Howe O, Devereux M (2011) Bis-phenanthroline copper(II) phthalate complexes are potent in vitro antitumour agents with 'self-activating' metallo-nuclease and DNA binding properties. Dalton Trans 40(5):1024–1027. doi:10.1039/c0dt01607a

60. Zhang M, Liu YQ, Ye BC (2011) Rapid and sensitive colorimetric visualization of phthalates using UTP-modified gold nanoparticles cross-linked by copper(ii). Chem Commun (Camb) 47(43):11849–11851. doi:10.1039/c1cc14772b

61. Wood CM, McDonald MD, Walker P, Grosell M, Barimo JF, Playle RC, Walsh PJ (2004) Bioavailability of silver and its relationship to ionoregulation and silver speciation across a range of salinities in the gulf toadfish (Opsanus beta). Aquat Toxicol 70(2):137–157. doi:10.1016/j.aquatox.2004.08.002

62. Boening DW (2000) Ecological effects, transport, and fate of mercury: a general review. Chemosphere 40(12):1335–1351. doi:10.1016/S0045-6535(99)00283-0

63. Ratte HT (1999) Bioaccumulation and toxicity of silver compounds: A review. Environ Toxicol Chem 18(1):89–108. doi:10.1002/etc.5620180112

64. Bhardwaj VK, Singh N, Hundal MS, Hundal G (2006) Mesitylene based azo-coupled chromogenic tripodal receptors–a visual detection of Ag(I) in aqueous medium. Tetrahedron 62(33):7878–7886. doi:10.1016/j.tet.2006.05.047

65. Harris HH, Pickering IJ, George GN (2003) The chemical form of mercury in fish. Science 301(5637):1203. doi:10.1126/science.1085941

66. Mutter J, Naumann J, Schneider R, Walach H, Haley B (2005) Mercury and autism: accelerating evidence? Neuro Endocrinol Lett 26(5):439–446. doi:NEL260505A10

67. Xue X, Wang F, Liu X (2008) One-step, room temperature, colorimetric detection of mercury (Hg^{2+}) using DNA/nanoparticle conjugates. J Am Chem Soc 130(11):3244–3245. doi:10. 1021/ja076716c

68. Li B, Du Y, Dong S (2009) DNA based gold nanoparticles colorimetric sensors for sensitive and selective detection of Ag(I) ions. Anal Chim Acta 644(1–2):78–82. doi:10.1016/ j.aca.2009.04.022

69. Thaxton CS, Georganopoulou DG, Mirkin CA (2006) Gold nanoparticle probes for the detection of nucleic acid targets. Clin Chim Acta 363(1–2):120–126. doi:10.1016/j.cccn. 2005.05.042

70. Ono A, Cao S, Togashi H, Tashiro M, Fujimoto T, Machinami T, Oda S, Miyake Y, Okamoto I, Tanaka Y (2008) Specific interactions between silver(I) ions and cytosine–cytosine pairs in DNA duplexes. Chem Commun (Camb) 39:4825–4827. doi:10.1039/b808686a

71. Chatterjee A, Santra M, Won N, Kim S, Kim JK, Kim SB, Ahn KH (2009) Selective fluorogenic and chromogenic probe for detection of silver ions and silver nanoparticles in aqueous media. J Am Chem Soc 131(6):2040–2041. doi:10.1021/ja807230c

72. Zhou XH, Kong DM, Shen HX (2010) G-quadruplex-hemin DNAzyme-amplified colorimetric detection of Ag + ion. Anal Chim Acta 678(1):124–127. doi:10.1016/j.aca. 2010.08.025

73. Wang H, Wang YX, Jin JY, Yang RH (2008) Gold nanoparticle-based colorimetric and "turn-on" fluorescent probe for mercury(II) ions in aqueous solution. Anal Chem 80(23):9021–9028. doi:10.1021/Ac801382k

74. Miao P, Liu L, Li Y, Li GX (2009) A novel electrochemical method to detect mercury (II) ions. Electrochem Commun 11(10):1904–1907. doi:10.1016/j.elecom.2009.08.013

75. Sanchez A, Walcarius A (2010) Surfactant-templated sol-gel silica thin films bearing 5-mercapto-1-methyl-tetrazole on carbon electrode for Hg(II) detection. Electrochim Acta 55(13):4201–4207. doi:10.1016/j.electacta.2010.02.016

76. Zhang M, Yin BC, Tan W, Ye BC (2011) A versatile graphene-based fluorescence "on/off" switch for multiplex detection of various targets. Biosens Bioelectron 26(7):3260–3265. doi:10.1016/j.bios.2010.12.037

Chapter 5
Metal-Nanoclusters-Based Luminescent Probe Design and Its Application

5.1 Introduction

Nobel metal nanoparticles, such as gold nanoparticles (AuNPs) and silver nanoparticles (AgNPs) (diameters greater than ~ 2 nm) have been the subject of intense research in the past decades due to their unique size- and shape-dependent catalytic, optical, electrical, magnetic, and chemical properties [1, 2]. The metal nanoparticles have proven to be of high utility in biochemical applications. When the particle size is further reduced to smaller than 2 nm, these ultra-small metal particles, called metal nanoclusters, can perform dramatically different optical, electrical and chemical properties compared to nanoparticles. MNCs, such as gold nanoclusters (AuNCs) and silver nanoclusters (AgNCs), normally consists of several to roughly hundred atoms. They possess sizes comparable to the Fermi wavelength of electrons and hence exhibit molecule-like properties, including discrete electronic transitions and strong fluorescence [3]. MNCs are like molecular species and size-dependent strong fluorescent emission can often be observed upon photoexcitation in the UV–visible range [4]. Very recently, there has been an explosion of interest in MNCs synthesis and their application in the area of biochemistry, especially AuNCs and AgNCs. Gold is well-known to be highly biocompatible and has been used in treatments of human inflammatory diseases for a long time [5]. Besides, AuNCs exhibit a potential for the targeted imaging of cancer in vitro and in vivo [6, 7]. To explore potential application of AuNCs, it is essential to synthesize stable, water-soluble individual AuNCs. Most AuNCs, protected by thiol-related compounds such as reduced glutathione (GSH), were prepared to emit fluorescence ranging from red to infrared (IR), but have low quantum yields [8–10]. In addition, other novel "green" synthetic routes for the preparation of red-emitting AuNCs used bovine serum albumin (BSA) [11–13], lysozyme [14], insulin [15], trypsin [16], or horseradish peroxidase (HRP) [17] etc., as the templates or capping agents. On the other hand, AgNCs have been reported to show brighter fluorescence than AuNCs in solutions, which have

B.-C. Ye et al., *Nano-Bio Probe Design and Its Application for Biochemical Analysis*, Springer Briefs in Molecular Science, DOI: 10.1007/978-3-642-29543-0_5, © The Author(s) 2012

received considerable attention in the past few years owing to their great promise
in a wide range of applications. Colloidal Ag is historically well-known to be
the core of most photographic processes. Fedrigo et al. firstly reported that gas-
phase and low-temperature matrix-isolated AgNCs can show discrete absorption
and fluorescence [18]. The synthesis of water-soluble fluorescent AgNCs was
reported by various methods, which were recently reviewed by Xu et al. [3].
Nowadays, metal nanoclusters-based analytical systems are a very active area with
the literature growing rapidly. In this chapter, we will review the recent
developments of metal-nanoclusters-based luminescent probe design and its
applications.

5.2 Metal-Nanoclusters-Based Luminescent Probes for Nucleic Acid Detection

There has been ever-increasing demand to develop rapid, sensitive, and selective
bioassays for the detection of DNA in a wide range of fields including genetics,
pathogenics, molecular diagnosis, antibioterrorism and drug development [19, 20].
A particularly attractive molecular tool toward this goal is the molecular beacon
(MB), which is a dually labeled hairpin-structured oligonucleotide which can be
internally quenched as a result of the close proximity between a fluorophore
(donor) and quencher (acceptor), tagged at either end. In the presence of a target
DNA, which is complementary to the loop of the hairpin structure, the MB
undergoes a conformational alteration upon hybridization from a closed (hairpin)
to an open (linear) structure, resulting in the separation of the fluorophore and the
quencher and the restoration of fluorescence. Traditional MBs have been employed
in a wide range of applications, but they have some limitations, such as false-
positive signals, insufficient sensitivity, high-cost synthesis, and difficult selection
of dye-quencher pair in certain cases [21]. Various hybridization-based DNA
detection modes also have been developed, such as chemiluminescent [22, 23],
electrochemistry [24, 25], and colorimetry [26, 27]. These methods all require
labeling the probe or target DNA molecules with a reporter. Therefore, to develop
the label-free detection mode is more favorable.

In the past decade, metal nanoclusters (MNCs) with few atoms, exhibiting size-
dependent fluorescence emission, have been developed as a new class of fluorophores, especially the synthesis of fluorescent AgNCs using DNA as templates in
aqueous solution has attracted extensive attention. The DNA-templated AgNCs
(DNA-AgNCs) exhibit outstanding spectral and photo-physical properties, which
are highly DNA sequence-dependent [28–30]. Richards et al. reported that the
photoluminescence (PL) emission band of DNA-AgNCs can be fine-tuned
throughout the visible and near infrared (NIR) range just by changing the sequence
of DNA [28]. These DNA-AgNCs hold great potential for many research fields,
especially in DNA-based nanomachines. Guo et al. recently demonstrated that
hybridized DNA duplexes can be used as capping scaffolds for the generation of

fluorescent AgNCs and the formation of fluorescent AgNCs is highly sequence-dependent. They have successfully applied these properties to identify the sickle cell anemia mutation, and furthermore, more general types of single nucleotide mismatches [31]. The highly sequence-dependent formation of fluorescent AgNCs holds great promise in DNA-hybridization-based analysis. Similarly, Lan et al. prepared fluorescent, functional oligonucleotide-stabilized AgNCs (FFDNA-AgNCs) through one-pot synthesis and then employed them as probes for DNA target and single nucleotide polymorphisms (SNPs) [32]. In their work, the FFDNA-AgNCs were obtained through the $NaBH_4$-mediated reduction of $AgNO_3$ in the presence of a DNA strand having the sequence $5'$-C_{12}-CCAGATACTCAC-CGG-$3'$. The specific DNA scaffold combines a fluorescent base motif (C_{12}) and a specific sequence (CCAGATACTCACCGG) that recognizes a gene for fumaryl-acetoacetate hydrolase (FAH). Yeh et al. reported that the red fluorescence of DNA-AgNCs can be enhanced 500-fold when placed in proximity to guanine (G)-rich DNA sequences [33]. On the basis of this new phenomenon, they have designed a DNA detection probe [NanoCluster Beacon (NCB)] that "lights up" upon DNA target binding.

MicroRNAs (miRNAs), 19–25 nt non-coding RNAs, play vital roles in numerous developmental, metabolic, and disease processes of plants and animals [34]. The individual levels of miRNAs can be useful biomarkers for cellular events or disease diagnosis, thus it is imperative to develop methods for sensitive and selective detection of miRNAs. Most recently, Yang and Vosch successfully developed a novel DNA-AgNC-based fluorescent probe for miRNAs detection [35]. They showed that the red fluorescence of the DNA-AgNC probe is diminished upon the presence of target miRNA without pre- or post-modification, addition of extra enhancer molecules, and labeling.

5.3 Metal-Nanoclusters-Based Luminescent Probes for Metal Ions Detection

Cu^{2+} plays primary role in many regulations of biological process. However, due to its widespread use, Cu^{2+} can also lead to serious environmental problems and is potentially toxic for all living organisms. Low-dose (<0.9 mg/day) of Cu^{2+} is an essential trace nutrient, but short-time exposure to high-dose of Cu^{2+} can cause gastrointestinal disturbance and long-time exposure can even cause damage to the liver and kidneys [36]. Many methods have been reported for Cu^{2+} detection, but most of them are time-consuming and impractical for a real-time or high-throughput format. Therefore, there is a high demand for a simple, rapid, sensitive, and high-throughput routine assay for Cu^{2+} sensing.

Nucleic acids are known to interact with metal ions through nucleobases and the phosphodiester backbone. Some researchers have revealed that Cu^{2+} can interact with uridine and its analogues [37]. Cu^{2+} is a well-known highly efficient fluorescent quencher due to its paramagnetic properties via electron or energy

Table 5.1 Sequences of DNA used in the work. Reprinted from Ref. [39] by permission of the Royal Society of Chemistry

Name	Sequence
T1	5′-ATCCTCCCACCGGGCCTCCCACCATAAAAACCCTTAATCCCC-3′
C1	5′-GCTTCTTTGTTGGGTTCTTTGTTGCAAAAACCCTTAATCCCC-3′
C2	5′-AGCGTAGAGACTGACCGTACTGTGCAAAAACCCTTAATCCCC-3′
C3	5′-CCCTTAATCCCC-3′

transfer [38]. Enlightened by the above facts, we think that a fluorescent assay for Cu^{2+} can be developed utilizing a designed DNA-templated AgNCs (DNA-AgNCs), and based on the fact that the interaction of nucleic acids and Cu^{2+} can quench the fluorescence of as-prepared AgNCs via metal–metal interplay. Herein, we presented a green chemical, simple and reliable fluorescent method for the quantitative determination of Cu^{2+} using DNA-AgNCs in a label-free format [39].

DNA-AgNCs were prepared by reducing $AgNO_3$ with sodium borohydride ($NaBH_4$) in the presence of a designed DNA template **T1**. The designed DNA template **T1** is a cytosine-(C)-rich ssDNA with the sequence: 5′-AT-CCTCCCACCGGGCCTCCCACCATAAAAACCCTTAATCCCC-3′, which was designed referring to the reported literature (Table 5.1) [40, 41]. Ag^+ is reported to selectively coordinate C bases and form stable C–Ag^+–C complexes [21, 42].

To synthesize **T1**–templated AgNCs (**T1**-AgNCs), **T1** and $AgNO_3$ were mixed in sodium phosphate buffer to form multiple C–Ag^+–C complexes. Following reduction with $NaBH_4$, the **T1**-AgNCs showed strong fluorescence. The absorption and fluorescence emissions were investigated to confirm the formation of AgNCs (Fig. 5.1a, b). The as-prepared **T1**-AgNCs showed their strong fluorescence compared to relatively low-fluorescent AgNCs using other control DNA (**C1**, **C2**, and **C3**).

As shown in Fig. 5.1c, the as-prepared DNA-AgNCs exhibit strong fluorescence emission at 624 nm with excitation at 564 nm. The DNA-AgNCs are highly dispersed in aqueous solution and become pink-red when illuminated by a UV lamp with excitation at 365 nm (Fig. 5.1c inset). The quantum yield of the DNA-AgNCs was calculated about 5.3% with Rhodamine 6G in ethanol as the reference at room temperature using the reported method [43, 44]. Photostability of DNA-AgNCs was investigated by illuminated samples with continuous excitation at 564 nm. The quantum dots (CdTe@CdS) were used as references. The results show DNA-AgNCs exhibit much less photobleaching than CdTe@CdS quantum dots. A typical transmission electron microscopy (TEM) image in Fig. 5.1d showed that the DNA-AgNCs are monodispersed and uniform with an average size of 1.8 nm. The DNA-AgNCs solution can be highly luminescent, however, in the presence of Cu^{2+}, the fluorescence of the DNA-AgNCs was found to be quenched by Cu^{2+}, which can be used as a selective "turn off" indicator for Cu^{2+} [39]. It is worth pointing out that Lan et al. described a turn-on and homogeneous assay—employing another DNA(**C3**)-stabilized AgNCs—for the highly selective and sensitive detection of Cu^{2+} [40]. The shorter DNA template **C3** is part of that of **T1**

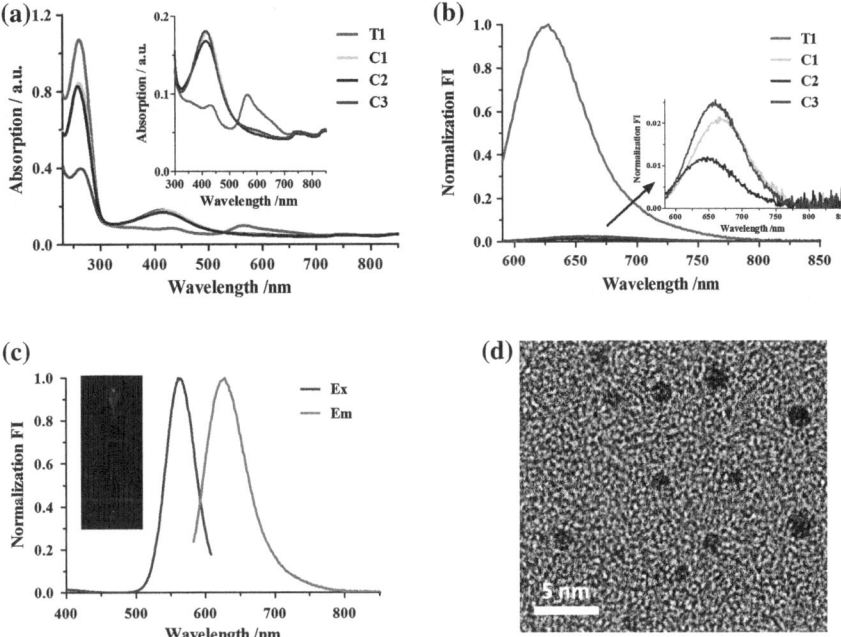

Fig. 5.1 **a** Typical absorption spectra of the fluorescent DNA-AgNCs (DNA template: **T1**, and other control DNA: **C1**, **C2**, and **C3**). Inset: magnification of the absorption spectra of DNA-AgNCs in the range of 300–850 nm. **b** The emission spectra of the fluorescent DNA-AgNCs (DNA template: **T1**, and other control DNA: **C1**, **C2**, and **C3**) excitation at 564 nm. Inset: magnification of the normalization fluorescence intensity (*FI*) of DNA-AgNCs (control DNA: **C1**, **C2**, and **C3**) in the range of 0–0.028. **c** The maximum excitation and emission spectra of the highly fluorescent DNA-AgNCs (Ex: 564 nm, Em: 624 nm). Inset: a photograph of the solution of DNA-AgNCs illuminated by a UV lamp with excitation at 365 nm. **d** A typical TEM image of the DNA-AgNCs. Reprinted from Ref. [39] by permission of the Royal Society of Chemistry

(Table 5.1). In their system, the fluorescent enhancement (an emission band centered at 564 nm) was attributed to the formation of the DNA-Cu/AgNCs to reveal a more rigid structure of the DNA template and result in the Cu-AgNCs being protected more completely by the DNA templates. However, in our work, the C-rich DNA template **T1** can capture more Ag^+ due to the formation of $C–Ag^+–C$ complexes, following reduction with $NaBH_4$, the resulting DNA-Ag-NCs showed strong red fluorescence, with an emission peak at 624 nm. In order to verify whether the quenching fluorescence of DNA-AgNCs was due to the reaction between Cu^{2+} and phosphate or base groups of oligonucleotide or the fluorescence quenching by metal–metal interaction, ethylenediaminetetraacetate (EDTA), a strong metal ion chelator, was used in competition with DNA-AgNCs for Cu^{2+}. The results showed that the chelation of Cu^{2+} by EDTA can slow down the response rate of fluoresce quenching but not to break off it, and the recovery of the fluorescence of DNA-AgNCs quenched by Cu^{2+} is unavailable (data not shown), which may confirm that the dominant factor of the quenching reaction is mainly the metal–metal interaction.

Considering the appreciable changes in fluorescent properties of DNA-AgNCs toward Cu^{2+}, the potential of developing a novel fluorescent probe for determination of Cu^{2+} was assessed. As shown in Fig. 5.2a, with the addition of an increasing concentration of Cu^{2+} to the suspension of DNA-AgNPs, an obvious decrease in the fluorescent peak at 624 nm (I_{624}) were clearly detected. From Fig. 5.2b, it can be seen that the fluorescence ratio (I_0/I) was sensitive to the concentration of Cu^{2+}, the fitting range is from 0 to 10 μM with an exponential growth equation $Y = 1.523*\exp(0.3264*X)$, where Y is the fluorescence ratio (I_0/I) and X is the concentration of Cu^{2+} (regression coefficient $R^2 = 0.9960$). Additionally, a liner equation can be obtained from the concentration range of 0–1 μM ($Y = 0.462 X + 1.009$, $R^2 = 0.9977$). The limit of detection of Cu^{2+} based on 3σ was approximately 0.01 μM, which has similar behavior to the PMMA-AgNCs method [45]. It is one of the most sensitive methods for the detection and analysis of Cu^{2+} in a label-free fluorescent format. To test selectivity, competing stimuli including different metal cations were examined under the same conditions as in the case of Cu^{2+} (Fig. 5.2c). It was found that Cu^{2+} resulted in an obvious change in the fluorescence, while there was nearly negligible fluorescent change in the presence of other stimuli. The results demonstrated the excellent selectivity of this approach applied in Cu^{2+} detection over other metal cations (Fig. 5.2d).

In addition to evaluating the Cu^{2+} sensor in a pure buffer artificial system, we further challenged the performance of the sensor in real world applications, i.e., in real environmental samples. We compared the time-course fluorescence quenching response of this assay upon reacting with blank river water and a river water sample spiked with Cu^{2+} ions (0.5, 1, 5 μM). Importantly, this assay was inert toward the blank sample, suggesting that the interference of materials in river water, such as bacteria and pathogens, could not quench the fluorescence of DNA-AgNCs probe. The fluorescence response of 0.5, 1, and 5 μM of Cu^{2+} ions was appreciably discriminable, which satisfied the sensitivity requirement (~ 20 μM) of U.S. Environmental Protection Agency (EPA) for drinking water, which indicated that the present method has a promise in practical application.

For the metal ions sensing, other research groups also reported varied assays based on fluorescent AgNCs or AuNCs. Tu et al. synthesized novel NIR luminescent reduced glutathione (GSH)-templated AuNCs (GSH-AuNCs) with emission peak maximum at 810 nm via a simple, rapid, and one-pot procedure, in which the driving force was attributed to the heat-assisted reduction of gold (I)-thiol complex. They demonstrated the sensing application for Cu^{2+} using these GSH-AuNCs [46]. Durgadas et al. reported the use of fluorescent AuNCs synthesized using BSA for the sensing of Cu^{2+} in live cells. The fluorescence of the BSA-AuNCs was found to be quenched by Cu^{2+}, enabling its detection in cells, and the "turn off" of fluorescence can be retrieved by a Cu^{2+} chelator glycine [47]. Su et al. developed a simple and homogeneous "turn on" fluorescence assay, comprised of 3-mercaptopropionic acid (MPA) and DNA-Cu/AgNCs in aqueous

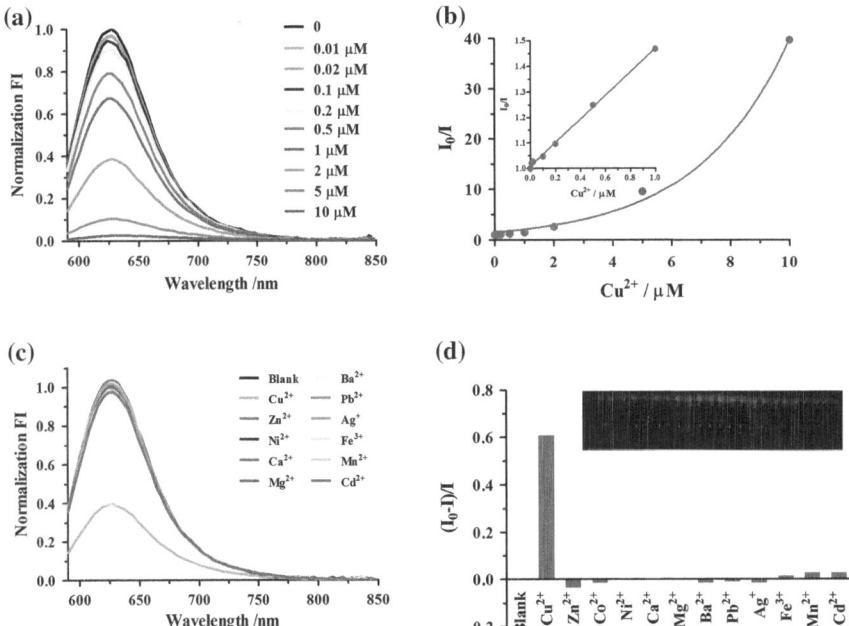

Fig. 5.2 **a** Emission spectra representing the quenching effect of different concentrations of Cu^{2+} on the fluorescent DNA-AgNCs; **b** Plot of fluorescence quenching effect (I_0/I) of the DNA-AgNCs at 624 nm incubated with various concentrations of Cu^{2+} (0, 0.01, 0.02, 0.1, 0.2, 0.5, 1, 2, 5, 10 μM). **c** Emission spectra representing the effect of 2 μM different tested metal ions toward DNA-AgNCs; **d** Fluorescence quenching effect [$(I_0-I)/I$] of the DNA-AgNCs at 624 nm incubated with a concentration of 2 μM different tested metal ions. Inset: photo of DNA-AgNCs was incubated with a concentration of 2 μM different tested metal ions illuminated by a UV lamp with excitation at 365 nm. Reprinted from Ref. [39] by permission of the Royal Society of Chemistry

solution, for the detection of Cu^{2+}. The fluorescence of the DNA-Cu/AgNCs was quenched by MPA, and was recovered in the presence of Cu^{2+} [48]. Liu et al. successfully prepared highly fluorescent and water-soluble AuNCs with NIR-emission and Au@AgNCs with yellow-emission via a rapid sonochemical approach, and the as-prepared AuNCs could be applied in the determination of Cu^{2+} [49].

Apart from Cu^{2+} sensing, some heavy metal ions like Hg^{2+} can also be detected via fluorescent metal nanoclusters. Ying and coworkers reported a Hg^{2+} sensor by using BSA-stabilized fluorescent AuNCs, the sensing mechanism was based on the high-affinity metallophilic $Hg^{2+}-Au^+$ interactions, which could effectively quench the fluorescence of AuNCs [50]. Wei et al. reported the preparation of lysozyme-stabilized fluorescent AuNCs by mixing lysozyme and $HAuCl_4$ under basic conditions, and its application as a Hg^{2+} sensor based on fluorescence quenching of AuNCs [14]. Kawasaki et al. reported on trypsin-stabilized fluorescent AuNCs for the sensitive and selective detection of Hg^{2+} [16]. Liu et al. reported a fast, robust

Table 5.2 Various types of sensors for metal ions made from fluorescent metal nanoclusters

Sensor materials	Template or capping agents	Synthesis method	Detection mode	Detected element	Detection limit	References
AgNCs	DNA	NaBH$_4$ reduction	Turn on	Cu^{2+}	8 nM	[32]
AgNCs	DNA	NaBH$_4$ reduction	Turn off	Cu^{2+}	10 nM	[31]
Cu/AgNCs	DNA	NaBH$_4$ reduction	Turn on	Cu^{2+}	2.7 nM	[41]
AgNCs	poly(methacrylic acid)	UV irradiation	Turn off	Cu^{2+}	8 nM	[38]
AuNCs	GSH	heat-assisted reduction	Turn off	Cu^{2+}	1.6 nM	[39]
AuNCs	BSA	basic conditions	Turn off	Cu^{2+}	50 μM	[40]
AuNCs	BSA	sonochemical approach	Turn off	Cu^{2+}	0.3 nM	[42]
AuNCs	BSA	basic conditions	Turn off	Hg^{2+}	0.5 nM	[43]
AuNCs	lysozyme	basic conditions	Turn off	Hg^{2+}	10 nM	[14]
AuNCs	trypsin	basic conditions	Turn off	Hg^{2+}	50 nM	[16]
AgNCs	poly(methacrylic acid)	microwave-assisted reduction	Turn off	Cr^{3+}	28 nM	[44]

microwave-assisted green synthesis of highly fluorescent AgNCs in the presence of a common polyelectrolyte, polymethacrylic acid sodium salt (PMAA-Na), and demonstrated the potential application for Cr^{3+} sensing based on the selective quenching effect of Cr^{3+} on the fluorescence emission of the AgNCs [51]. Table 5.2 summaries the recently reported sensors for metal ions made from fluorescent metal nanoclusters.

5.4 Metal-Nanoclusters-Based Luminescent Probes for Cellular Labeling or Imaging

The advances in fluorescence microscopy facilitate the better understanding of biological systems, numerous organic dyes with unique photo-physical properties have been utilized as indicators for tracking processes in living cells [52]. It is pointed out that imaging intracellular dynamics with organic dyes is hampered by brightness, photostability, and strategies for specific labeling, especially in single molecule studies [53]. The emergence and marvelous development of nanotechnology and nanoscience provides great opportunity for the cellular labeling or imaging researches, which utilize novel nanomaterials with unique photo-physical properties. Quantum dots (QDs), semiconductor fluorescent nanocrystals (diameters ∼2–100 nm) typically contain a CdSe or CdTe core and ZnS shell. They have been found to be superior to traditional organic dyes on several counts, such as high quantum yield, narrow, symmetric and stable fluorescence, and size-dependent and tunable absorption and emission [54]. However, the intrinsic toxicity of QDs does harm to human subjects due to QD's cadmium release [55], and self-aggregation of QDs that occurred in the living cells fatally limit their practical biomedical applications [56]. Another type of fluorophore frequently used is fluorescent proteins, which has attracted attention as fluorescent labels in the last decade [57]. A revolution in cellular imaging has resulted from the discovery and development of the green fluorescent protein (GFP) from the jellyfish *Aequorea victoria*, which was reviewed by Tsien [58]. Extensive researches were conducted to refine the fluorescent properties of GFP from *Aequorea victoria* [59, 60], GFP mutants, and related proteins from other marine species [61–63]. However, fluorescent proteins may suffer from moderate brightness and fast blinking/photoleaching [64, 65]. Recently, DNA-templated AgNCs have emerged as a novel class of bright and photostable fluorophores, which can exhibit fine-tunable emission spectra from blue to NIR wavelengths [28, 29, 66]. For in vivo imaging, the use of red and NIR emitters is preferable, because there is a major drawback in using UV and blue light. Several abundant biomolecules in cells show high absorption in UV and blue wavelength ranges, thereby reducing light penetration through tissue, moreover, many intracellular species exhibit weak fluorescence when illuminated at these wavelengths, thereby increasing background [53]. Antoku et al. recently reported that the NIR emitting $C_{24}:Ag_n$ can be introduced into living HeLa cells by using lipofectamine as a transfection agent. Bright NIR fluorescence was observed from inside the transfected HeLa cells, when exciting with 633 nm excitation, opening up the possibility for the use of these $C_{24}:Ag_n$ clusters for biological labeling and imaging of living cells and for monitoring the transfection process with limited harm to the living cells [53]. Furthermore, Yu et al. reported that DNA-encapsulated AgNCs are readily conjugated to proteins and serve as alternatives to organic dyes and semiconductor QDs. Stable and bright on the bulk and single molecule levels, AgNCs fluorescence is readily observed when staining live cell surfaces [67].

Apart from AgNCs, AuNCs also can be used as fluorescent probes for cellular imaging, because the fluorescent AuNCs, with ultrafine sizes, do not disturb the biological functions of the labeled bioentities [68]. Shang et al. recently developed a facile strategy to synthesize water-soluble fluorescent AuNCs capped with bidentate ligand dihydrolipoic acid (DHLA) [69]. They demonstrated the DHLA-capped AuNCs possess many attractive features including ultra-small size, bright NIR luminescence, high colloidal stability, good biocompatibility, and long fluorescence lifetime (>100 ns), which make them promising imaging agents for biomedical and cellular imaging applications, as an example, the internalization of AuNCs by live HeLa cells was visualized using the fluorescence lifetime imaging technique. Lin et al. successfully used water-soluble fluorescent AuNCs capped with DHLA and modified with polyethylene glycol (PEG), BSA, and streptavidin for cell imaging [70]. Retnakumari et al. demonstrated the potential of using non-toxic fluorescent BSA-AuNCs conjugated to folic acid for the targeted imaging of cancer [6]. Similarly, Wu et al. reported that the fluorescent BSA-AuNCs can be applied in MDA-MB-45 and HeLa tumor xenograft model imaging [7]. Liu et al. reported the synthesis of fluorescent AuNCs by using insulin as a template. They demonstrated that the resulting insulin-AuNCs exhibit intense red fluorescence, retain their bioactivity and biocompatibility and show versatility in applications such as fluorescence imaging, CT, and in vivo blood-glucose regulation [15]. Wang et al. developed a novel biocompatible marker for in vitro and in vivo tracking of endothelial cells utilizing fluorescent AuNCs, and the results showed that the fluorescent AuNCs-labeled cells not only exhibited a strong fluorescence but also maintained intact angiogenic potential [71].

References

1. Park J, Joo J, Kwon SG, Jang Y, Hyeon T (2007) Synthesis of monodisperse spherical nanocrystals. Angew Chem Int Ed Engl 46(25):4630–4660. doi:10.1002/anie.200603148
2. Xia YN, Xiong YJ, Lim B, Skrabalak SE (2009) Shape-controlled synthesis of metal nanocrystals: simple chemistry meets complex physics? Angew Chem Int Edit 48(1):60–103. doi:10.1002/anie.200802248
3. Xu H, Suslick KS (2010) Water-soluble fluorescent silver nanoclusters. Adv Mater 22(10):1078–1082. doi:10.1002/adma.200904199
4. Zheng J, Nicovich PR, Dickson RM (2007) Highly fluorescent noble-metal quantum dots. Annu Rev Phys Chem 58:409–431. doi:10.1146/annurev.physchem.58.032806.104546
5. Lehman AJ, Esdaile JM, Klinkhoff AV, Grant E, Fitzgerald A, Canvin J (2005) A 48-week, randomized, double-blind, double-observer, placebo-controlled multicenter trial of combination methotrexate and intramuscular gold therapy in rheumatoid arthritis: results of the METGO study. Arthritis Rheum 52(5):1360–1370. doi:10.1002/art.21018
6. Retnakumari A, Setua S, Menon D, Ravindran P, Muhammed H, Pradeep T, Nair S, Koyakutty M (2010) Molecular-receptor-specific, non-toxic, near-infrared-emitting Au cluster-protein nanoconjugates for targeted cancer imaging. Nanotechnology 21(5):055103. doi:10.1088/0957-4484/21/5/055103
7. Wu X, He X, Wang K, Xie C, Zhou B, Qing Z (2010) Ultrasmall near-infrared gold nanoclusters for tumor fluorescence imaging in vivo. Nanoscale 2(10):2244–2249. doi:10.1039/c0nr00359j
8. Negishi Y, Takasugi Y, Sato S, Yao H, Kimura K, Tsukuda T (2004) Magic-numbered Au(n) clusters protected by glutathione monolayers (n = 18, 21, 25, 28, 32, 39): isolation and spectroscopic characterization. J Am Chem Soc 126(21):6518–6519. doi:10.1021/ja0483589
9. Wang G, Huang T, Murray RW, Menard L, Nuzzo RG (2005) Near-IR luminescence of monolayer-protected metal clusters. J Am Chem Soc 127(3):812–813. doi:10.1021/ja0452471
10. Chen CT, Chen WJ, Liu CZ, Chang LY, Chen YC (2009) Glutathione-bound gold nanoclusters for selective-binding and detection of glutathione S-transferase-fusion proteins from cell lysates. Chem Commun (Camb) 48:7515–7517. doi:10.1039/b916919a
11. Hu D, Sheng Z, Gong P, Zhang P, Cai L (2010) Highly selective fluorescent sensors for Hg(2+) based on bovine serum albumin-capped gold nanoclusters. Analyst 135(6): 1411–1416. doi:10.1039/c000589d
12. Li HW, Ai K, Wu Y (2011) Fluorescence visual gel-separation of dansylated BSA-protected gold-nanoclusters. Chem Commun (Camb) 47(35):9852–9854. doi:10.1039/c1cc12588e
13. Xie J, Zheng Y, Ying JY (2009) Protein-directed synthesis of highly fluorescent gold nanoclusters. J Am Chem Soc 131(3):888–889. doi:10.1021/ja806804u
14. Wei H, Wang Z, Yang L, Tian S, Hou C, Lu Y (2010) Lysozyme-stabilized gold fluorescent cluster: Synthesis and application as Hg(2 +) sensor. Analyst 135(6):1406–1410. doi:10.1039/c0an00046a
15. Liu CL, Wu HT, Hsiao YH, Lai CW, Shih CW, Peng YK, Tang KC, Chang HW, Chien YC, Hsiao JK, Cheng JT, Chou PT (2011) Insulin-directed synthesis of fluorescent gold

nanoclusters: preservation of insulin bioactivity and versatility in cell imaging. Angew Chem Int Ed Engl 50(31):7056–7060. doi:10.1002/anie.201100299

16. Kawasaki H, Yoshimura K, Hamaguchi K, Arakawa R (2011) Trypsin-stabilized fluorescent gold nanocluster for sensitive and selective Hg^{2+} detection. Anal Sci 27(6):591–596. doi: JST.JSTAGE/analsci/27.591

17. Wen F, Dong Y, Feng L, Wang S, Zhang S, Zhang X (2011) Horseradish peroxidase functionalized fluorescent gold nanoclusters for hydrogen peroxide sensing. Anal Chem 83(4):1193–1196. doi:10.1021/ac1031447

18. Fedrigo S, Harbich W, Buttet J (1993) Collective dipole oscillations in small silver clusters embedded in rare-gas matrices. Phys Rev B Condens Matter 47(16):10706–10715. doi:10.1103/PhysRevB.47.10706

19. Debouck C, Goodfellow PN (1999) DNA microarrays in drug discovery and development. Nat Genet 21(1 Suppl):48–50. doi:10.1038/4475

20. Heller MJ (2002) DNA microarray technology: devices, systems, and applications. Annu Rev Biomed Eng 4:129–153. doi:10.1146/annurev.bioeng.4.020702.153438

21. Zhang M, Yin BC, Tan W, Ye BC (2011) A versatile graphene-based fluorescence "on/off" switch for multiplex detection of various targets. Biosens Bioelectron 26(7):3260–3265. doi:10.1016/j.bios.2010.12.037

22. Li H, He Z (2009) Magnetic bead-based DNA hybridization assay with chemiluminescence and chemiluminescent imaging detection. Analyst 134(4):800–804. doi:10.1039/b819990f

23. Fan A, Lau C, Lu J (2009) Hydroxylamine-amplified gold nanoparticles for the naked eye and chemiluminescent detection of sequence-specific DNA with notable potential for single-nucleotide polymorphism discrimination. Analyst 134(3):497–503. doi:10.1039/b817047a

24. Zhang D, Peng Y, Qi H, Gao Q, Zhang C (2010) Label-free electrochemical DNA biosensor array for simultaneous detection of the HIV-1 and HIV-2 oligonucleotides incorporating different hairpin-DNA probes and redox indicator. Biosens Bioelectron 25(5):1088–1094. doi:10.1016/j.bios.2009.09.032

25. Xiao Y, Lou X, Uzawa T, Plakos KJ, Plaxco KW, Soh HT (2009) An electrochemical sensor for single nucleotide polymorphism detection in serum based on a triple-stem DNA probe. J Am Chem Soc 131(42):15311–15316. doi:10.1021/ja905068s

26. Kanjanawarut R, Su X (2009) Colorimetric detection of DNA using unmodified metallic nanoparticles and peptide nucleic acid probes. Anal Chem 81(15):6122–6129. doi:10.1021/ac900525k

27. Xu W, Xue X, Li T, Zeng H, Liu X (2009) Ultrasensitive and selective colorimetric DNA detection by nicking endonuclease assisted nanoparticle amplification. Angew Chem Int Ed Engl 48(37):6849–6852. doi:10.1002/anie.200901772

28. Richards CI, Choi S, Hsiang JC, Antoku Y, Vosch T, Bongiorno A, Tzeng YL, Dickson RM (2008) Oligonucleotide-stabilized Ag nanocluster fluorophores. J Am Chem Soc 130(15):5038–5039. doi:10.1021/ja8005644

29. O'Neill PR, Velazquez LR, Dunn DG, Gwinn EG, Fygenson DK (2009) Hairpins with Poly-C loops stabilize four types of fluorescent Ag(n):DNA. J Phys Chem C 113(11):4229–4233. doi:10.1021/Jp809274m

30. Gwinn EG, O'Neill P, Guerrero AJ, Bouwmeester D, Fygenson DK (2008) Sequence-dependent fluorescence of DNA-hosted silver nanoclusters. Adv Mater 20 (2):279. doi:10.1002/adma.200702380

31. Guo W, Yuan J, Dong Q, Wang E (2010) Highly sequence-dependent formation of fluorescent silver nanoclusters in hybridized DNA duplexes for single nucleotide mutation identification. J Am Chem Soc 132(3):932–934. doi:10.1021/ja907075s

32. Lan GY, Chen WY, Chang HT (2011) One-pot synthesis of fluorescent oligonucleotide Ag nanoclusters for specific and sensitive detection of DNA. Biosens Bioelectron 26(5):2431–2435. doi:10.1016/j.bios.2010.10.026

33. Yeh HC, Sharma J, Han JJ, Martinez JS, Werner JH (2010) A DNA-silver nanocluster probe that fluoresces upon hybridization. Nano Lett 10(8):3106–3110. doi:10.1021/nl101773c

34. Buchan JR, Parker R (2007) Molecular biology. The two faces of miRNA. Science 318(5858):1877–1878. doi:10.1126/science.1152623
35. Yang SW, Vosch T (2011) Rapid detection of MicroRNA by a silver nanocluster DNA probe. Anal Chem 83(18):6935–6939. doi:10.1021/ac201903n
36. Georgopoulos PG, Roy A, Yonone-Lioy MJ, Opiekun RE, Lioy PJ (2001) Environmental copper: its dynamics and human exposure issues. J Toxicol Environ Health B Crit Rev 4(4):341–394. doi:10.1080/109374001753146207
37. Lomozik L, Jastrzab R (2003) Copper(II) complexes with uridine, uridine 5′-monophosphate, spermidine, or spermine in aqueous solution. J Inorg Biochem 93(3–4):132–140. doi:S0162013402005676
38. Varnes AW, Dodson RB, Wehry EL (1972) Interactions of transition-metal ions with photoexcited states of flavins. Fluorescence quenching studies. J Am Chem Soc 94(3): 946–950
39. Zhang M, Ye BC (2011) Label-free fluorescent detection of copper(ii) using DNA-templated highly luminescent silver nanoclusters. Analyst 136(24):5139–5142. doi:10.1039/c1an15891k
40. Lan GY, Huang CC, Chang HT (2010) Silver nanoclusters as fluorescent probes for selective and sensitive detection of copper ions. Chem Commun (Camb) 46(8):1257–1259. doi:10.1039/b920783j
41. Sharma J, Yeh HC, Yoo H, Werner JH, Martinez JS (2011) Silver nanocluster aptamers: in situ generation of intrinsically fluorescent recognition ligands for protein detection. Chem Commun (Camb) 47(8):2294–2296. doi:10.1039/c0cc03711g
42. Huy GD, Zhang M, Zuo P, Ye BC (2011) Multiplexed analysis of silver(I) and mercury(II) ions using oligonucletide-metal nanoparticle conjugates. Analyst 136(16):3289–3294. doi:10.1039/c1an15373k
43. Zhong X, Feng Y, Knoll W, Han M (2003) Alloyed Zn(x)Cd(1-x)S nanocrystals with highly narrow luminescence spectral width. J Am Chem Soc 125(44):13559–13563. doi:10.1021/ja036683a
44. Zhong X, Han M, Dong Z, White TJ, Knoll W (2003) Composition-tunable Zn(x)Cd(1-x)Se nanocrystals with high luminescence and stability. J Am Chem Soc 125(28):8589–8594. doi:10.1021/ja035096m
45. Shang L, Dong SJ (2008) Silver nanocluster-based fluorescent sensors for sensitive detection of Cu(II). J Mater Chem 18(39):4636–4640. doi:10.1039/B810409c
46. Tu X, Chen W, Guo X (2011) Facile one-pot synthesis of near-infrared luminescent gold nanoparticles for sensing copper (II). Nanotechnology 22(9):095701. doi:10.1088/0957-4484/22/9/095701
47. Durgadas CV, Sharma CP, Sreenivasan K (2011) Fluorescent gold clusters as nanosensors for copper ions in live cells. Analyst 136(5):933–940. doi:10.1039/C0an00424c
48. Su YT, Lan GY, Chen WY, Chang HT (2010) Detection of copper ions through recovery of the fluorescence of DNA-templated copper/silver nanoclusters in the presence of mercaptopropionic acid. Anal Chem 82(20):8566–8572. doi:10.1021/Ac101659d
49. Liu H, Zhang X, Wu X, Jiang L, Burda C, Zhu JJ (2011) Rapid sonochemical synthesis of highly luminescent non-toxic AuNCs and Au@AgNCs and Cu (II) sensing. Chem Commun (Camb) 47(14):4237–4239. doi:10.1039/c1cc00103e
50. Xie JP, Zheng YG, Ying JY (2010) Highly selective and ultrasensitive detection of Hg(2+) based on fluorescence quenching of Au nanoclusters by Hg(2+)-Au(+) interactions. Chem Commun 46(6):961–963. doi:10.1039/B920748a
51. Liu S, Lu F, Zhu JJ (2011) Highly fluorescent Ag nanoclusters: microwave-assisted green synthesis and Cr3+ sensing. Chem Commun (Camb) 47(9):2661–2663. doi:10.1039/c0cc04276e
52. Giepmans BN, Adams SR, Ellisman MH, Tsien RY (2006) The fluorescent toolbox for assessing protein location and function. Science 312(5771):217–224. doi:10.1126/science.1124618

53. Antoku Y, Hotta J, Mizuno H, Dickson RM, Hofkens J, Vosch T (2010) Transfection of living HeLa cells with fluorescent poly-cytosine encapsulated Ag nanoclusters. Photochem Photobiol Sci 9(5):716–721. doi:10.1039/c0pp00015a

54. Michalet X, Pinaud FF, Bentolila LA, Tsay JM, Doose S, Li JJ, Sundaresan G, Wu AM, Gambhir SS, Weiss S (2005) Quantum dots for live cells, in vivo imaging, and diagnostics. Science 307(5709):538–544. doi:10.1126/science.1104274

55. Gao XH, Yang LL, Petros JA, Marshal FF, Simons JW, Nie SM (2005) In vivo molecular and cellular imaging with quantum dots. Curr Opin Biotech 16(1):63–72. doi:10.1016/j.copbio.2004.11.003

56. Weng JF, Ren JC (2006) Luminescent quantum dots: a very attractive and promising tool in biomedicine. Curr Med Chem 13(8):897–909

57. Blum C, Subramaniam V (2009) Single-molecule spectroscopy of fluorescent proteins. Anal Bioanal Chem 393(2):527–541. doi:10.1007/s00216-008-2425-x

58. Tsien RY (1998) The green fluorescent protein. Annu Rev Biochem 67:509–544. doi:10.1146/annurev.biochem.67.1.509

59. Zimmer M (2009) GFP: from jellyfish to the Nobel prize and beyond. Chem Soc Rev 38(10):2823–2832. doi:10.1039/b904023d

60. Seward HE, Bagshaw CR (2009) The photochemistry of fluorescent proteins: implications for their biological applications. Chem Soc Rev 38(10):2842–2851. doi:10.1039/b901355p

61. Cotlet M, Hofkens J, Habuchi S, Dirix G, Van Guyse M, Michiels J, Vanderleyden J, De Schryver FC (2001) Identification of different emitting species in the red fluorescent protein DsRed by means of ensemble and single-molecule spectroscopy. Proc Natl Acad Sci U S A 98(25):14398–14403. doi:10.1073/pnas.251532698

62. Betzig E, Patterson GH, Sougrat R, Lindwasser OW, Olenych S, Bonifacino JS, Davidson MW, Lippincott-Schwartz J, Hess HF (2006) Imaging intracellular fluorescent proteins at nanometer resolution. Science 313(5793):1642–1645. doi:10.1126/science.1127344

63. Mocz G (2007) Fluorescent proteins and their use in marine biosciences, biotechnology, and proteomics. Mar Biotechnol (NY) 9(3):305–328. doi:10.1007/s10126-006-7145-7

64. Flors C, Hotta J, Uji-i H, Dedecker P, Ando R, Mizuno H, Miyawaki A, Hofkens J (2007) A stroboscopic approach for fast photoactivation-localization microscopy with Dronpa mutants. J Am Chem Soc 129(45):13970–13977. doi:10.1021/ja074704l

65. Folling J, Bossi M, Bock H, Medda R, Wurm CA, Hein B, Jakobs S, Eggeling C, Hell SW (2008) Fluorescence nanoscopy by ground-state depletion and single-molecule return. Nat Methods 5(11):943–945. doi:10.1038/nmeth.1257

66. Vosch T, Antoku Y, Hsiang JC, Richards CI, Gonzalez JI, Dickson RM (2007) Strongly emissive individual DNA-encapsulated Ag nanoclusters as single-molecule fluorophores. Proc Natl Acad Sci U S A 104(31):12616–12621. doi:10.1073/pnas.0610677104

67. Yu J, Choi S, Richards CI, Antoku Y, Dickson RM (2008) Live cell surface labeling with fluorescent Ag nanocluster conjugates. Photochem Photobiol 84(6):1435–1439. doi:10.1111/j.1751-1097.2008.00434.x

68. Liu CL, Ho ML, Chen YC, Hsieh CC, Lin YC, Wang YH, Yang MJ, Duan HS, Chen BS, Lee JF, Hsiao JK, Chou PT (2009) Thiol-functionalized gold nanodots: two-photon absorption property and imaging in vitro. J Phys Chem C 113(50):21082–21089. doi:10.1021/Jp9080492

69. Shang L, Azadfar N, Stockmar F, Send W, Trouillet V, Bruns M, Gerthsen D, Nienhaus GU (2011) One-pot synthesis of near-infrared fluorescent gold clusters for cellular fluorescence lifetime imaging. Small 7(18):2614–2620. doi:10.1002/smll.201100746

70. Lin CA, Yang TY, Lee CH, Huang SH, Sperling RA, Zanella M, Li JK, Shen JL, Wang HH, Yeh HI, Parak WJ, Chang WH (2009) Synthesis, characterization, and bioconjugation of fluorescent gold nanoclusters toward biological labeling applications. ACS Nano 3(2):395–401. doi:10.1021/nn800632j

71. Wang HH, Lin CAJ, Lee CH, Lin YC, Tseng YM, Hsieh CL, Chen CH, Tsai CH, Hsieh CT, Shen JL, Chan WH, Chang WH, Yeh HI (2011) Fluorescent gold nanoclusters as a biocompatible marker for in vitro and in vivo tracking of endothelial cells. ACS Nano 5(6):4337–4344. doi:10.1021/Nn102752a